Fuzzy Logic

Edited by Constantin Volosencu

Published in London, United Kingdom

IntechOpen

Supporting open minds since 2005

Fuzzy Logic
http://dx.doi.org/10.5772/intechopen.77460
Edited by Constantin Volosencu

Part of IntechOpen Book Series: Artificial Intelligence, Volume 3
Book Series Editor: Marco Antonio Aceves-Fernandez

Contributors
Mohammed Bsiss, Benaissa Amami, Venkata Subba Reddy Poli, Dmitrii Nazarov, Sy Dzung Nguyen, Oleksandr Rubanenko, Lezhniuk Petro D., Rubanenko Olena O., Vakeel Ahmad Khan, Hira Fatima, Mobeen Ahmad, Mustafa Babanli, Constantin Volosencu

First published in London, United Kingdom, 2020 by IntechOpen
IntechOpen is the global imprint of INTECHOPEN LIMITED, registered in England and Wales, registration number: 11086078, 7th floor, 10 Lower Thames Street, London, EC3R 6AF, United Kingdom
Printed in Croatia

British Library Cataloguing-in-Publication Data
A catalogue record for this book is available from the British Library

Additional hard and PDF copies can be obtained from orders@intechopen.com

Fuzzy Logic
Edited by Constantin Volosencu
p. cm.
Print ISBN 978-1-78984-231-9
Online ISBN 978-1-78984-232-6
eBook (PDF) ISBN 978-1-83968-540-8
ISSN 2633-1403

For EU product safety concerns:
IN TECH d.o.o., Prolaz Marije Krucifikse Kozulić 3, 51000 Rijeka, Croatia, info@intechopen.com or visit our website at intechopen.com.

IntechOpen Book Series

Artificial Intelligence

Volume 3

Constantin Volosencu is a professor at the "Politehnica" University of Timisoara, Department of Automation. He is the author of 10 books, four book chapters, more than 150 scientific papers published in journals and conference proceedings, and 27 patents. He is the editor of seven books, and manager of research grants. He is a member of the editorial boards of international journals, a former plenary speaker, a member of several scientific committees, and chair at international conferences. He research is in the fields of control systems, electrical drives, power ultrasounds, fuzzy logic, neural networks, fault detection and diagnosis, sensor networks, and distributed parameter systems. He has developed electrical equipment for machine tools, spooling machines, high-power ultrasound processes and others, with homologation of 18 prototypes and 12 zero manufacturing series.

Editor of Volume 3:
Constantin Volosencu
"Politehnica" University from Timisoara, Romania

Book Series Editor: Marco A. Aceves-Fernandez
Universidad Autonoma de Queretaro, Mexico

Scope of the Series

Artificial Intelligence (AI) is a rapidly developing multidisciplinary research area that aims to solve increasingly complex problems. In today´s highly integrated world, AI promises to become a robust and powerful mean for obtaining solutions to previously unsolvable problems. This book series is intended for researchers and students alike, as well as all those interested in this fascinating field and its applications, in particular in areas related to the topics on which it is focused.

Contents

Preface

This book promotes new research results in the field of advanced fuzzy logic applications. Treating information using fuzzy logic has developed over the past 50 years, and this mathematical theory is an interesting tool for researchers to solve complex scientific and technical problems. Fuzzy logic has found applications in various sectors of human activity, such as: industry, business, finance, medicine, and more. The book includes new research results in scientific fields such as: fuzzy mathematics, adaptive neuro-fuzzy systems, inference methods, fuzzy control, expert systems, dynamic fuzzy neural networks, and others. The authors have published worked examples and case studies resulting from their research in the field. Readers will have access to new solutions and answers to questions related to emerging theoretical fuzzy logic applications and their implementation.

The book has eight sections: Introduction, Fuzzy Mathematics, Adaptive Neuro-Fuzzy Inference System, Inference Methods, Expert Systems, Control of Electrical Systems, Fuzzy Logic Applications in Management, and Field-Programmable Gate Array for Fuzzy Controllers.

The book includes an introductory chapter that presents basic properties of fuzzy relations and seven main chapters that illustrate research in the section domains.

The chapters were edited and published following a rigorous selection process, with only a small number of the proposed chapters being selected for publication.

The introductory chapter aims to recall algebraic relations that describe fuzzy rule bases and fuzzy blocks as algebraic applications. Also, the fuzzy block may be described graph-analytically with transfer functions and graphs.

The second chapter includes a study on the convergence of sequence spaces with respect to intuitionistic fuzzy norms and their topological and algebraic properties. The third chapter focuses mainly on building ANFIS and its application to identifying the online bearing fault. A traditional structure of ANFIS as a data-driven model is shown. A recurrent mechanism depicting the relation between the processes of filtering impulse noise and establishing ANFIS from a noisy measuring database is presented. One of the typical applications of ANFIS related to managing online bearing fault is presented. The fourth chapter presents methods of conditional inference for fuzzy control systems. The fifth chapter presents an application of fuzzy logic and fuzzy expert systems in material synthesis methods. The data-driven approach is used to construct a fuzzy model. Fuzzy C-means clustering is used to derive fuzzy if-then rules from material data that describe material composition. The sixth chapter presents an example of how to use fuzzy logic in control of electrical systems, in conditions of incomplete information regarding the values of diagnostic parameters. The seventh chapter includes a fuzzy logic methodology for evaluating the causality of factors in organization management. The chapter formulates the problem of causal relations in a broad sense and analyzes the methods for its solution with an emphasis on socioeconomic aspects. Systems approach, comparative experiment, economic and mathematical modeling, and other general scientific methods are used. The eighth chapter includes a technical study on functional safety of an FPGA fuzzy logic controller. This chapter proposes to analyze the implementation of advanced safety architecture of fuzzy logic controllers with 1-out-of-2 controllers in FPGA using the reliability block diagram and the Markov model.

The editor thanks the authors for their excellent contributions in the field and their understanding during the process of editing. Also, the editor thanks all the editorial personnel involved in book publication. The publishing process provided a set of editorial standards, which ensured the quality of the scientific level of relevance of accepted chapters.

Constantin Volosencu
Professor,
"Politehnica" University from Timisoara,
Romania

Section 1

Introduction

Chapter 1

Introductory Chapter: Basic Properties of Fuzzy Relations

Constantin Volosencu

1. General aspects

Treating information using fuzzy logic has developed over the past 50 years, this mathematical theory being an interesting tool for researchers to solve complex scientific and technical problems. In these years, research has always yielded new results in the field of advanced fuzzy logic applications. Fuzzy logic has found applications in various sectors of human activity, such as, industry, business, finance, medicine, and in many scientific fields such as, machine learning, big data technologies, fuzzy control, expert systems, dynamic fuzzy neural networks, and others. Fuzzy logic provides a different way of dealing with mathematical calculus problems. In the case of fuzzy logic, conventional algorithms are replaced by a series of linguistic rules of the If (then) condition (conclusion). Thus, a heuristic algorithm is obtained, and human experience can be taken into account in the subject matter of the calculation.

This introductory chapter aims to recall some basic notions, main properties of fuzzy relations. Fuzzy rule bases and fuzzy blocks may be seen as relations between fuzzy sets and, respectively, between real sets, with algebraic properties as commutative property, inverse and identity. The fuzzy relations are developed with different rule bases, fuzzy values, membership functions, inference, and defuzzification methods, and they may be characterized with transfer characteristic graphs.

The book [1] can be considered a reference in the field. Other references may be taken in consideration [2–5]. The author published also in the field [6].

As advantages of fuzzy logic are useful in the calculations, we can list the following:

- Development of fuzzy controllers without a complex mathematical modeling of the problem addressed
- The possibility of implementing “human linguistic knowledge” on the application solved.
- The possibility of using fuzzy logic for complex, nonlinear, and variable relations
- The possibility of performing exceptional treatments, that is, changing the calculation strategies as a result of a change in the course of application
- Possibility of using it when making decisions specific to artificial intelligence
- Interpolation among rules, usable in exceptional treatments to change the scope of application

A fuzzy relation is a composed relation of defuzzification, inference, based on rules, and fuzzification. In the development of fuzzy relations, we have to answer the following questions. The lack of precise directives for conceiving a fuzzy relation. And in this case the following questions will be answered:

- What is the structure of the fuzzy relation?
- What real mathematical variables have to be chosen for fuzzy processing?
- How to choose the universes of discourse for fuzzy variables?
- How many fuzzy values and what membership functions are chosen for fuzzy relationship variables?
- Which is the rule base of the fuzzy relation?
- Which method of inference should be chosen?
- Which defuzzification method is better?

To answer these questions, a large number of fuzzy relations have been experimented, and calculus tests have been performed in a comparative analysis. The answers to these questions are reported by the values of the desired efficiency indicators and the values that can be provided by each fuzzy relation variant. By answering the questions posed, the empirical and unsystematic character of the operator's knowledge implementation and the synthesis of the fuzzy logic-based relation can be eliminated at a later design.

Next, for a better understanding of the phenomena occurring in the fuzzy relations, a brief presentation of their main basic properties will be made.

2. Properties

2.1 Fuzzy relation

The basic fuzzy relation is a function of two variables:

$$y = f(x_1, x_2) \tag{1}$$

The variables are defined on universes of discourse, as real sets:

$$x_1 \in X_1, x_2 \in X_2, y \in Y \tag{2}$$

A fuzzy relation may be described informationally by a structure as in **Figure 1**. It is composed relation, from defuzzification, inference based on rules, and fuzzification. The fuzzy values are defined and described with membership functions, defined on universes of discourse, with values on interval [0, 1]:

$$m(x): X \rightarrow [0,1] \tag{3}$$

The fuzzy set is defined as

$$A = \{x, m(x)\} \tag{4}$$

The membership functions are represented as graphs. A fuzzy variable with three fuzzy values NB, ZE, and PB and also three membership functions is represented in **Figure 2**.

A fuzzy variable may have also five or seven fuzzy values.

The fuzzy relation is developed based on a rule base, for a fuzzy reasoning of the form [If x_1 is ... and x_2 is ... then y is ...]. A primary rule base of 3 × 3 rules is presented in **Table 1**.

Several inference methods may be use, for example, max-min and sum-prod.

Also there are some defuzzification methods: center of gravity, mean of maxima, and others.

An example of inference max-min is presented in **Figure 3**.

2.2 Algebraic properties

The fuzzy relations have the following algebraic properties.

Commutative property

$$f(x_1,x_2) = f(x_2,x_1) \tag{5}$$

Inverse of x is $-x$:

$$f(x,-x) = f(-x,x) = 0 \tag{6}$$

Identity is 0 (ZE):

$$f(x,0) = f(0,x) = x \tag{7}$$

The rule bases have also the same properties.

But they do not have the associative property and nor the property of distributivity.

2.3 Graphs

The fuzzy relation is characterized by some graphs [6].

First is the graph of function (1), represented in **Figure 4a**.

The second graph is the graph of y with x_1 as variable and x_2 as parameter:

$$y = f(x_1;x_2) \tag{8}$$

represented as a family of characteristics in **Figure 4b**. The third graph is a family of characteristics:

$$y = f(x_t;x_2) \tag{9}$$

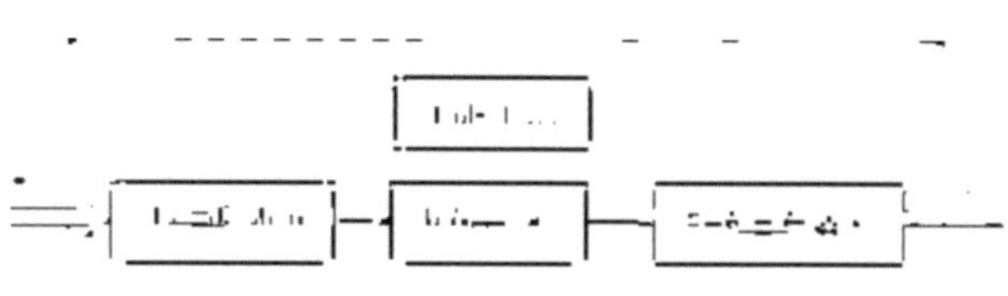

Figure 1.
The structure of a fuzzy relation.

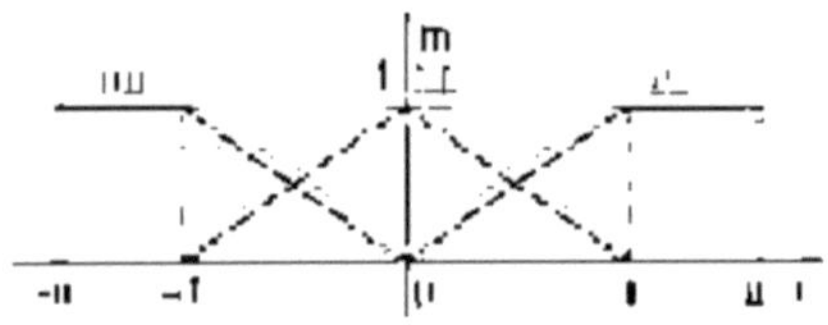

Figure 2.
Membership function.

y		x1		
		NB	**ZE**	**PB**
x2	NB	NB	NB	ZE
	ZE	NB	ZE	PB
	PB	ZE	PB	PB

Table 1.
Primary 3 × 3 rule base.

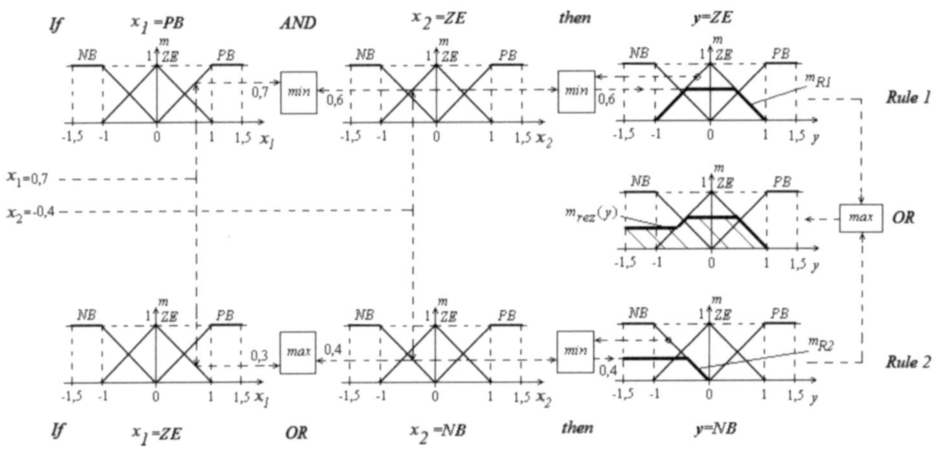

Figure 3.
Example of inference max-min.

represented in **Figure 4c**, where

$$x_t = x_1 + x_2 \tag{10}$$

is a compound variable. This graph is situated in the first and third quadrants and it has a sector property.

And the fourth graph is the variable gain:

$$K(x_t;x_2) = \frac{y}{x_t} \tag{11}$$

represented, as a family of characteristics, in **Figure 4d**, with the value in origin:

$$K_0 = \lim_{x_t \to 0} \frac{y}{x_t} \tag{12}$$

The graphs are obtained for a fuzzy relation with three fuzzy values, membership function from **Figure 2**, the primary 3 × 3 rule base, max-min inference, and defuzzification with center of gravity.

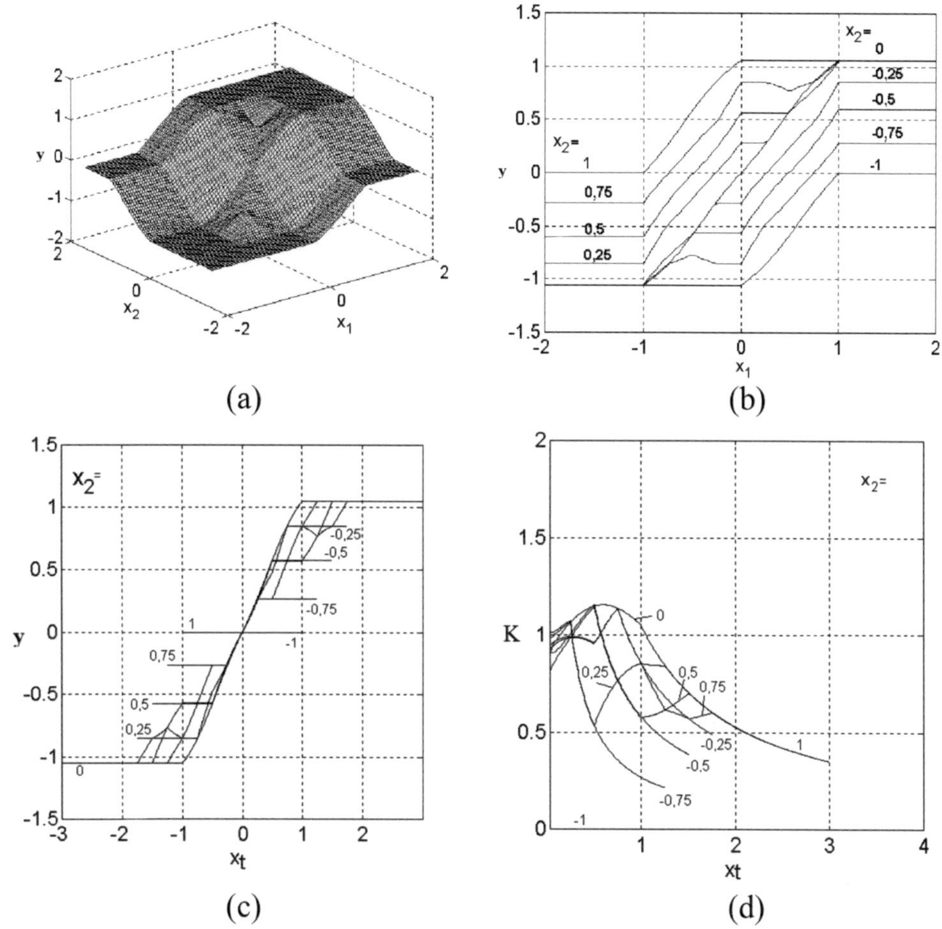

Figure 4.
Graphs of a fuzzy relation.

3. Conclusion

The fuzzy relations may be classified according the rule base, membership functions, number of fuzzy variables, inference, and defuzzification. They have transfer characteristic graphs which may be numerical calculated. The graphs may be used for grapho-analytical analysis of fuzzy relations and their applications, because only the analytical description of the fuzzy systems is difficult because of the complexity of operations made inside: fuzzification, inference, and defuzzification. The rule bases and the fuzzy relations may have algebraic properties, the commutative property, inverse, and identity, but not the associative property, so no kind of algebraic structures may be developed. The fuzzy relations are nonlinear functions. They have applications in many domain, like fuzzy controllers with variable gain, for example.

Author details

Constantin Volosencu
"Politehnica" University, Timisoara, Romania

*Address all correspondence to: constantin.volosencu@aut.upt.ro

References

[1] Buhler H. Reglage par Logique Floue. Lausanne: Press Polytechnique et Universitaires Romands; 1994

[2] Cox E. Adaptive fuzzy systems. IEEE Spectrum. 1993

[3] Mendel JM. Fuzzy logic systems for engineering: A tutorial. Proceedings of the IEEE. 1995

[4] Thomas DE, Armstrong-Helouvry B. Fuzzy logic control, a taxonomy of demonstrated benefits. Proceedings of the IEEE. 1995

[5] Zimmerman HJ. Fuzzy Sets Theory and its Applications. Boston: Kluwer-Nijhoff Pub; 1985

[6] Volosencu C. Properties of fuzzy systems. WSEAS Transactions on Systems. 2009;**8**(2):210-228

Section 2

Fuzzy Mathematics

Chapter 2

Some Topological Properties of Intuitionistic Fuzzy Normed Spaces

Vakeel Ahmad Khan, Hira Fatima and Mobeen Ahmad

Abstract

In 1986, Atanassov introduced the concept of intuitionistic fuzzy set theory which is based on the extensions of definitions of fuzzy set theory given by Zadeh. This theory provides a variable model to elaborate uncertainty and vagueness involved in decision making problems. In this chapter, we concentrate our study on the ideal convergence of sequence spaces with respect to intuitionistic fuzzy norm and discussed their topological and algebraic properties.

Keywords: ideal, intuitionistic fuzzy normed spaces, Orlicz function compact operator, I-convergence

1. Introduction

In recent years, the fuzzy theory has emerged as the most active area of research in many branches of mathematics, computer and engineering [1]. After the excellent work of Zadeh [2], a large number of research work have been done on fuzzy set theory and its applications as well as fuzzy analogues of the classical theories. It has a wide number of applications in various fields such as population dynamics [3], nonlinear dynamical system [4], chaos control [5], computer programming [6], etc. In 2006, Saadati and Park [7] introduced the concept of intuitionistic fuzzy normed spaces after that the concept of statistical convergence in intuitionistic fuzzy normed space was studied for single sequence in [8]. The study of intuitionistic fuzzy topological spaces [9], intuitionistic fuzzy 2-normed space [10] and intuitionistic fuzzy Zweier ideal convergent sequence spaces [11] are the latest developments in fuzzy topology.

First, let us recall some notions, basic definitions and concepts which are used in sequel.

Definition 1.1. (See Ref. [7]). The five-tuple $(X, \mu, \nu, *, \diamond)$ is said to be an intuitionistic fuzzy normed space (for short, IFNS) if X is a vector space, $*$ is a continuous t-norm, $\diamond$ is a continuous t-conorm, and μ and ν are fuzzy sets on $X \times (0, \infty)$ satisfying the following conditions for every $x, y \in X$ and $s, t > 0$:

(a) $\mu(x,t) + \nu(x,t) \leq 1$,

(b) $\mu(x,t) > 0$,

(c) $\mu(x,t) = 1$ if and only if $x = 0$,

(d) $\mu(\alpha x,t) = \mu\left(x, \frac{t}{|\alpha|}\right)$ for each $\alpha \neq 0$,

(e) $\mu(x,t) * \mu(y,s) \leq \mu(x+y,t+s)$,
(f) $\mu(x,.) : (0,\infty) \to [0,1]$ is continuous,
(g) $\lim_{t\to\infty} \mu(x,t) = 1$ and $\lim_{t\to 0} \mu(x,t) = 0$,
(h) $\nu(x,t) < 1$,
(i) $\nu(x,t) = 0$ if and only if $x = 0$,
(j) $\nu(\alpha x,t) = \nu\left(x, \frac{t}{|\alpha|}\right)$ for each $\alpha \neq 0$,
(k) $\nu(x,t) \diamond \nu(y,s) \geq \nu(x+y,t+s)$,
(l) $\nu(x,.) : (0,\infty) \to [0,1]$ is continuous,
(m) $\lim_{t\to\infty} \nu(x,t) = 0$ and $\lim_{t\to 0} \nu(x,t) = 1$.

In this case (μ,ν) is called an intuitionistic fuzzy norm.

Example 1.1. *Let* $(X, \|.\|)$ *be a normed space. Denote* $a * b = ab$ *and* $a \diamond b = \min(a+b,1)$ *for all* $a, b \in [0,1]$ *and let* μ_0 *and* ν_0 *be fuzzy sets on* $X \times (0,\infty)$ *defined as follows:*

$$\mu_0(x,t) = \frac{t}{t+\|x\|}, \quad \text{and} \quad \nu_0(x,t) = \frac{\|x\|}{t+\|x\|}$$

for all $t \in \mathbb{R}^+$. Then $(X,\mu,\nu,*,\diamond)$ is an intuitionistic fuzzy normed space.

Definition 1.2. Let $(X,\mu,\nu,*,\diamond)$ be an IFNS. Then a sequence $x = (x_k)$ is said to be convergent to $L \in X$ with respect to the intuitionistic fuzzy norm (μ,ν) if, for every $\varepsilon > 0$ and $t > 0$, there exists $k_0 \in \mathbb{N}$ such that $\mu(x_k - L,t) > 1 - \varepsilon$ and $\nu(x_k - L,t) < \varepsilon$ for all $k \geq k_0$. In this case we write (μ,ν)-$\lim x = L$.

In 1951, the concept of statistical convergence was introduced by Steinhaus [12] and Fast [13] in their papers "Sur la convergence ordinaire et la convergence asymptotique" and "Sur la convergence statistique," respectively. Later on, in 1959, Schoenberg [14] reintroduced this concept. It is a very useful functional tool for studying the convergence of numerical problems through the concept of density. The concept of ideal convergence, which is a generalization of statistical convergence, was introduced by Kostyrko et al. [15] and it is based on the ideal I as a subsets of the set of positive integers and further studied in [16–20].

Let X be a non-empty set then a family $I \subset 2^X$ is said to be an **ideal** in X if $\emptyset \in I$, I is additive, i.e., for all $A, B \in I \Rightarrow A \cup B \in I$ and I is hereditary, i.e., for all $A \in I$, $B \subseteq A \Rightarrow B \in I$. A non empty family of sets $\mathcal{F} \subset 2^X$ is said to be a **filter** on X if for all $A, B \in \mathcal{F}$ implies $A \cap B \in \mathcal{F}$ and for all $A \in \mathcal{F}$ with $A \subseteq B$ implies $B \in \mathcal{F}$. An ideal $I \subset 2^X$ is said to be **nontrivial** if $I \neq 2^X$, this non trivial ideal is said to be admissible if $I \supseteq \{\{x\} : x \in X\}$ and is said to be **maximal** if there cannot exist any nontrivial ideal $J \neq I$ containing I as a subset. For each ideal I, there is a filter $\mathcal{F}(I)$ called as filter associate with ideal I, that is (see [15]),

$$\mathcal{F}(I) = \{K \subseteq X : K^c \in I\}, \quad \text{where} \quad K^c = X \backslash K. \tag{1}$$

A sequence $x = (x_k) \in \omega$ is said to be I-**convergent** [21, 22] to a number L if for every $\varepsilon > 0$, we have $\{k \in \mathbb{N} : |x_k - L| \geq \varepsilon\} \in I$. In this case, we write $I - \lim_{xk} = L$.

2. IF-ideal convergent sequence spaces using compact operator

This section consists of some double sequence spaces with respect to intuitionistic fuzzy normed space and study the fuzzy topology on the said spaces. First we recall some basic definitions on compact operator.

Definition 2.1. (See [23]). Let X and Y be two normed linear spaces and $T : \mathcal{D}(T) \to Y$ be a linear operator, where $\mathcal{D} \subset X$. Then, the operator T is said to be bounded, if there exists a positive real k such that

$$\|Tx\| \leq k\|x\|, \quad \text{for all} \quad x \in \mathcal{D}(T).$$

The set of all bounded linear operators $\mathcal{B}(X, Y)$ [24] is a normed linear spaces normed by

$$\|T\| = \sup_{x \in X, \ \|x\|=1} \|Tx\|$$

and $\mathcal{B}(X, Y)$ is a Banach space if Y is a Banach space.

Definition 2.2. (See [23]). Let X and Y be two normed linear spaces. An operator $T : X \to Y$ is said to be a compact linear operator (or completely continuous linear operator), if

(i) T is linear,

(ii) T maps every bounded sequence (x_k) in X on to a sequence $(T(x_k))$ in Y which has a convergent subsequence.

The set of all compact linear operators $\mathcal{C}(X, Y)$ is a closed subspace of $\mathcal{B}(X, Y)$ and $\mathcal{C}(X, Y)$ is Banach space, if Y is a Banach space.

In 2015, Khan et al. [11] introduced the following sequence spaces:

$$\mathcal{Z}^I_{(\mu,\nu)} = \left\{(x_k) \in \omega : \left\{k : \mu\left(x'_k - L, t\right) \leq 1 - \varepsilon \ \ or \ \ \nu\left(x'_k - L, t\right) \geq \varepsilon\right\} \in I\right\},$$
$$\mathcal{Z}^I_{0(\mu,\nu)} = \left\{(x_k) \in \omega : \left\{k : \mu\left(x'_k, t\right) \leq 1 - \varepsilon \ \ or \ \ \nu\left(x'_k, t\right) \geq \varepsilon\right\} \in I\right\}.$$

Motivated by this, we introduce the following sequence spaces with the help of compact operator in intuitionistic fuzzy normed spaces:

$$\mathcal{M}^I_{(\mu,\nu)}(T) = (x_k) \in \ell_\infty : \{k : \mu(T(x_k) - L, t) \leq 1 - \varepsilon \ \text{or} \ \nu(T(x_k) - L, t) \geq \varepsilon \in I\} \tag{2}$$

$$\mathcal{M}^I_{0(\mu,\nu)}(T) = (x_k) \in \ell_\infty : \{k : \mu(T(x_k), t) \leq 1 - \varepsilon \ \text{or} \ \nu(T(x_k), t) \geq \varepsilon \in I\}. \tag{3}$$

Here, we also define an open ball with center x and radius r with respect to t as follows:

$$\mathcal{B}_x(r, t)(T) = (y_k) \in \ell_\infty : \{k : \mu(T(x_k) - T(y_k), t) \leq 1 - \varepsilon \ \text{or} \ \nu(T(x_k) - T(y_k), t) \geq \varepsilon \in I\}. \tag{4}$$

Now, we are ready to state and prove our main results. This theorem is based on the linearity of new define sequence spaces which is stated as follows.

Theorem 2.1. The sequence spaces $\mathcal{M}^I_{(\mu,\nu)}(T)$ and $\mathcal{M}^I_{0(\mu,\nu)}(T)$ are linear spaces.

Proof. Let $x = (x_k), y = (y_k) \in \mathcal{M}^I_{(\mu,\nu)}(T)$ and α, β be scalars. Then for a given $\varepsilon > 0$, we have the sets:

$$P_1 = \left\{k : \mu\left(T(x_k) - L_1, \frac{t}{2|\alpha|}\right) \leq 1 - \varepsilon \ \text{or} \ \nu\left(T(x_k) - L_1, \frac{t}{2|\alpha|}\right) \geq \varepsilon\right\} \in I;$$
$$P_2 = \left\{k : \mu\left(T(y_k) - L_2, \frac{t}{2|\beta|}\right) \leq 1 - \varepsilon \ \text{or} \ \nu\left(T(y_k) - L_2, \frac{t}{2|\beta|}\right) \geq \varepsilon\right\} \in I.$$

This implies

$$P_1^c = \left\{k : \mu\left(T(x_k) - L_1, \frac{t}{2|\alpha|}\right) > 1 - \varepsilon \ \text{or} \ \nu\left(T(x_k) - L_1, \frac{t}{2|\alpha|}\right) < \varepsilon\right\} \in \mathcal{F}(I);$$

$$P_2^c = \left\{k : \mu\left(T\left(_{yk}\right) - L_2, \frac{t}{2|\beta|}\right) > 1 - \varepsilon \text{ or } \nu\left(\left(T\left(_{yk}\right) - L_2, \frac{t}{2|\beta|}\right) < \varepsilon\right\} \in \mathcal{F}(I).$$

Now, we define the set $P_3 = P_1 \cup P_2$, so that $P_3 \in I$. It shows that P_3^c is a non-empty set in $\mathcal{F}(I)$. We shall show that for each (x_k), $(y_k) \in \mathcal{M}^I_{(\mu,\nu)}(T)$.

$$P_3^c \subset \{k : \mu((\alpha T(x_k) + \beta T(y_k)) - (\alpha L_1 + \beta L_2), t) > 1 - \varepsilon$$
$$\text{or } \nu((\alpha T(x_k) + \beta T(y_k)) - (\alpha L_1 + \beta L_2), t) < \varepsilon\}.$$

Let $m \in P_3^c$, in this case

$$\mu\left(T(x_m) - L_1, \frac{t}{2|\alpha|}\right) > 1 - \varepsilon \text{ or } \nu\left(T(x_m) - L_1, \frac{t}{2|\alpha|}\right) < \varepsilon$$

and

$$\mu\left(T(y_m) - L_2, \frac{t}{2|\beta|}\right) > 1 - \varepsilon \text{ } or \text{ } \nu\left(T(y_m) - L_2, \frac{t}{2|\beta|}\right) < \varepsilon.$$

Thus, we have

$$\begin{aligned} &\mu((\alpha T(x_m) + \beta T(y_m)) - (\alpha L_1 + \beta L_2), t) \\ &\geq \mu\left(\alpha T(x_m) - \alpha L_1, \frac{t}{2}\right) * \mu\left(\beta T(x_m) - \beta L_2, \frac{t}{2}\right) \\ &= \mu\left(T(x_m) - L_1, \frac{t}{2|\alpha|}\right) * \mu\left(T(x_m) - L_2, \frac{t}{2|\beta|}\right) \\ &> (1 - \varepsilon) * (1 - \varepsilon) = 1 - \varepsilon. \end{aligned}$$

and

$$\begin{aligned} &\nu((\alpha T(x_m) + \beta T(y_m)) - (\alpha L_1 + \beta L_2), t) \\ &\leq \nu\left(\alpha T(x_m) - \alpha L_1, \frac{t}{2}\right) \diamond \nu\left(\beta T(x_m) - \beta L_2, \frac{t}{2}\right) \\ &= \mu\left(T(x_m) - L_1, \frac{t}{2|\alpha|}\right) \diamond \mu\left(T(x_m) - L_2, \frac{t}{2|\beta|}\right) \\ &< \varepsilon \diamond \varepsilon = \varepsilon. \end{aligned}$$

This implies that

$$P_3^c \subset \{k : \mu((\alpha T(x_k) + \beta T(y_k)) - (\alpha L_1 + \beta L_2), t) > 1 - \varepsilon$$
$$\text{or } \nu((\alpha T(x_k) + \beta T(y_k)) - (\alpha L_1 + \beta L_2), t) < \varepsilon.$$

Therefore, the sequence space $\mathcal{M}^I_{(\mu,\nu)}(T)$ is a linear space.

Similarly, we can proof for the other space. □

In the following theorems, we discussed the convergence problem in the said sequence spaces. For this, firstly we have to discuss about the topology of this space. Define

$$\tau^I_{(\mu,\nu)}(T) = A \subset \mathcal{M}^I_{(\mu,\nu)}(T) : \text{for each } x \in A \text{ there exists } t > 0$$
$$\text{and } r \in (0,1) \text{ such that } \mathcal{B}_x(r,t)(T) \subset A.$$

Then $\tau^I_{(\mu,\nu)}(T)$ is a topology on $\mathcal{M}^I_{(\mu,\nu)}(T)$.

Theorem 2.2. Let $\mathcal{M}^I_{(\mu,\nu)}(T)$ is an IFNS and $\tau^I_{(\mu,\nu)}(T)$ is a topology on $\mathcal{M}^I_{(\mu,\nu)}(T)$. Then a sequence $(x_k)\in\mathcal{M}^I_{(\mu,\nu)}(T)$, $x_k \to x$ if and only if $\mu(T(x_k) - T(x), t) \to 1$ and $\nu(T(x_k) - T(x), t) \to 0$ as $k \to \infty$.

Proof. Fix $t_0>0$. Suppose $x_k \to x$. Then for $r\in(0,1)$, there exists $n_0\in\mathbb{N}$ such that $(x_k)\in\mathcal{B}_x(r,t)(T)$ for all $k\geq n_0$. So, we have

$$\mathcal{B}_x(r,t_0)(T) = \{k : \mu(T(x_k) - T(x), t)\leq 1 - r \text{ or } \nu(T(x_k) - T(x), t_0)\geq r\}\in I,$$

such that $\mathcal{B}^c_x(r,t_0)(T)\in\mathcal{F}(I)$. Then $1 - \mu(T(x_k) - T(x), t_0) < r$ and $\nu(T(x_k) - T(x), t_0) < r$. Hence $\mu(T(x_k) - T(x), t_0) \to 1$ and $\nu(T(x_k) - T(x), t_0) \to 0$ as $k \to \infty$.

Conversely, if for each $t>0$, $\mu(T(x_k) - T(x), t) \to 1$ and $\nu(T(x_k) - T(x), t) \to 0$ as $k \to \infty$, then for $r\in(0,1)$, there exists $n_0\in\mathbb{N}$, such that $1 - \mu(T(x_k) - T(x), t) < r$ and $\nu(T(x_k) - T(x), t) < r$, for all $k\geq n_0$. It shows that $\mu(T(x_k) - T(x), t)>1 - r$ and $\nu(T(x_k) - T(x), t) < r$ for all $k\geq n_0$. Therefore $(x_k)\in\mathcal{B}^c_x(r,t)(T)$ for all $k\geq n_0$ and hence $x_k \to x$.

There are some facts that arise in connection with the convergence of sequences in these spaces. Let us proceed to the next theorem on Ideal convergence of sequences in these new define spaces.

Theorem 2.3. A sequence $x = (x_k)\in\mathcal{M}^I_{(\mu,\nu)}(T)$ is I-convergent if and only if for every $\varepsilon>0$ and $t>0$ there exists a number $N = N(x,\varepsilon,t)$ such that

$$\left\{N : \mu\left(T(x_N) - L, \frac{t}{2}\right)>1 - \varepsilon \text{ or } \nu\left(T(x_N) - L, \frac{t}{2}\right) < \varepsilon\right\}\in\mathcal{F}(I).$$

Proof. Suppose that $I_{(\mu,\nu)} - \lim x = L$ and let $t>0$. For a given $\varepsilon>0$, choose $s>0$ such that $(1-\varepsilon)*(1-\varepsilon)>1 - s$ and $\varepsilon\diamond\varepsilon < s$. Then for each $x\in\mathcal{M}^I_{(\mu,\nu)}(T)$,

$$R = \left\{k : \mu\left(T(x_k) - L, \frac{t}{2}\right)\leq 1 - \varepsilon \text{ or } \nu\left(T(x_k) - L, \frac{t}{2}\right)\geq\varepsilon\right\}\in I,$$

which implies that

$$R^c = \left\{k : \mu\left(T(x_k) - L, \frac{t}{2}\right)>1 - \varepsilon \text{ or } \nu\left(T(x_k) - L, \frac{t}{2}\right) < \varepsilon\right\}\in\mathcal{F}(I).$$

Conversely, let us choose $N\in R^c$. Then

$$\mu\left(T(x_N) - L, \frac{t}{2}\right)>1 - \varepsilon \;\; or \;\; \nu\left(T(x_N) - L, \frac{t}{2}\right) < \varepsilon.$$

Now, we want to show that there exists a number $N = N(x,\varepsilon,t)$ such that

$$\{k : \mu(T(x_k) - T(x_N), t)\leq 1 - s \text{ or } \nu(T(x_k) - T(x_N), t)\geq s\}\in I.$$

For this, we define for each $x\in\mathcal{M}^I_{(\mu,\nu)}(T)$

$$S = \{k : \mu(T(x_k) - T(x_N), t)\leq 1 - s \text{ or } \nu(T(x_k) - T(x_N), t)\geq s\}\in I.$$

So, we have to show that $S\subset R$. Let us suppose that $S\subsetneq R$, then there exists $n\in S$ and $n\notin R$. Therefore, we have

$$\mu(T(x_n) - T(x_N), t)\leq 1 - s \text{ or } \mu\left(T(x_n) - L, \frac{t}{2}\right)>1 - \varepsilon.$$

In particular $\mu\left(T(x_N) - L, \frac{t}{2}\right) > 1 - \varepsilon$. Therefore, we have

$$1 - s \geq \mu(T(x_n) - T(x_N), t) \geq \mu\left(T(x_n) - L, \frac{t}{2}\right) * \mu\left(T(x_N) - L, \frac{t}{2}\right) \geq (1 - \varepsilon) * (1 - \varepsilon) > 1 - s,$$

which is not possible. On the other hand

$$\nu(T(x_n) - T(x_N), t) \geq s \ \ or \ \ \nu\left(T(x_n) - L, \frac{t}{2}\right) < \varepsilon.$$

In particular $\nu\left(T(x_N) - L, \frac{t}{2}\right) < \varepsilon$. So, we have

$$s \leq \nu(T(x_n) - T(x_N), t) \leq \nu\left(T(x_n) - L, \frac{t}{2}\right) \diamond \nu\left(T(x_N) - L, \frac{t}{2}\right) \leq \varepsilon \diamond \varepsilon < s,$$

which is not possible. Hence $S \subset R$. $R \in I$ which implies $S \in I$. □

3. IF-ideal convergent sequence spaces using Orlicz function

In this section, we have discussed the ideal convergence of sequences in Intuitionistic fuzzy I-convergent sequence spaces defined by compact operator and Orlicz function. We shall now define the concept of Orlicz function, which is basic definition in our work.

Definition 3.1. An Orlicz function is a function $F : [0, \infty) \to [0, \infty)$, which is continuous, non-decreasing and convex with $F(0) = 0$, $F(x) > 0$ for $x > 0$ and $F(x) \to \infty$ as $x \to \infty$. If the convexity of Orlicz function F is replaced by $F(x + y) \leq F(x) + F(y)$, then this function is called modulus function.

Remark 3.1. If F is an Orlicz function, then $F(\lambda x) \leq \lambda F(x)$ for all λ with $0 < \lambda < 1$.

In 2009, Mohiuddine and Lohani [18] introduced the concept of statistical convergence in intuitionistic fuzzy normed spaces in their paper published in Chaos, Solitons and Fractals. This motivated us to introduced some sequence spaces defined by compact operator and Orlicz function which are as follows:

$$\mathcal{M}^I_{(\mu,\nu)}(T, F) = \left\{ (x_k) \in \ell_\infty : \{k : F\left(\frac{\mu(T(x_k) - L, t)}{\rho}\right) \leq 1 - \varepsilon \right.$$
$$\left. \text{or } F\left(\frac{\nu(T(x_k) - L, t)}{\rho}\right) \geq \varepsilon\} \in I \right\}; \tag{5}$$

$$\mathcal{M}^I_{0(\mu,\nu)}(T, F) = \left\{ (x_k) \in \ell_\infty : \{k : F\left(\frac{\mu(T(x_k), t)}{\rho}\right) \leq 1 - \varepsilon \right.$$
$$\left. \text{or } F\left(\frac{\nu(T(x_k), t)}{\rho}\right) \geq \varepsilon\} \in I \right\}. \tag{6}$$

We also define an open ball with center x and radius r with respect to t as follows:

$$\mathcal{B}_x(r, t)(T, F) = \left\{ (y_k) \in \ell_\infty : k : F\left(\frac{\mu(T(x_k) - T(y_k), t)}{\rho}\right) \leq 1 - \varepsilon \right.$$
$$\left. \text{or } F\left(\frac{\nu(T(x_k) - T(y_k), t)}{\rho}\right) \geq \varepsilon \in I \right\}. \tag{7}$$

We shall now consider some theorems of these sequence spaces and invite the reader to verify the linearity of these sequence spaces.

Theorem 3.1. Every open ball $\mathcal{B}_x(r, t)(T, F)$ is an open set in $\mathcal{M}^I_{(\mu,\nu)}(T, F)$.

Proof. Let $\mathcal{B}_x(r,t)(T,F)$ be an open ball with center x and radius r with respect to t. That is

$$\mathcal{B}_x(r,t)(T,F) = \left\{ y = (y_k) \in \ell_\infty : \left\{ k : F\left(\frac{\mu(T(x_k) - T(y_k), t)}{\rho}\right) \leq 1 - r \right.\right.$$
$$\left.\left. \text{or } F\left(\frac{\nu(T(x_k) - T(y_k), t)}{\rho}\right) \geq r \right\} \in I \right\}.$$

Let $y \in \mathcal{B}_x^c(r,t)(T,F)$, then $F\left(\frac{\mu(T(x_k) - T(y_k), t)}{\rho}\right) > 1 - r$ and $F\left(\frac{\nu(T(x_k) - T(y_k), t)}{\rho}\right) < r$. Since $F\left(\frac{\mu(T(x_k) - T(y_k), t)}{\rho}\right) > 1 - r$, there exists $t_0 \in (0,t)$ such that $F\left(\frac{\mu(T(x_k) - T(y_k), t_0)}{\rho}\right) > 1 - r$ and $F\left(\frac{\nu(T(x_k) - T(y_k), t_0)}{\rho}\right) < r$.

Putting $r_0 = F\left(\frac{\mu(T(x_k) - T(y_k), t_0)}{\rho}\right)$, so we have $r_0 > 1 - r$, there exists $s \in (0,1)$ such that $r_0 > 1 - s > 1 - r$. For $r_0 > 1 - s$, we have $r_1, r_2 \in (0,1)$ such that $r_0 * r_1 > 1 - s$ and $(1 - r_0) \diamond (1 - r_0) \leq s$. Putting $r_3 = \max\{r_1, r_2\}$. Now we consider a ball $\mathcal{B}_y^c(1 - r_3, t - t_0)(T,F)$. And we prove that

$$\mathcal{B}_y^c(1 - r_3, t - t_0)(T,F) \subset \mathcal{B}_x^c(r,t)(T,F).$$

Let $z = (z_k) \in \mathcal{B}_y^c(1 - r_3, t - t_0)(T,F)$, then $F\left(\frac{\mu(T(y_k) - T(z_k), t - t_0)}{\rho}\right) > r_3$ and $F\left(\frac{\nu(T(y_k) - T(z_k), t - t_0)}{\rho}\right) < 1 - r_3$. Therefore, we have

$$F\left(\frac{\mu(T(x_k) - T(z_k), t)}{\rho}\right) \geq F\left(\frac{\mu(T(x_k) - T(y_k), t_0)}{\rho}\right) * F\left(\frac{\mu(T(y_k) - T(z_k), t - t_0)}{\rho}\right)$$
$$\geq (r_0 * r_3) \geq (r_0 * r_1) \geq (1 - s) \geq (1 - r)$$

and

$$F\left(\frac{\nu(T(x_k) - T(z_k), t)}{\rho}\right) \leq F\left(\frac{\nu(T(x_k) - T(y_k), t_0)}{\rho}\right) \diamond F\left(\frac{\nu(T(y_k) - T(z_k), t - t_0)}{\rho}\right)$$
$$\leq (1 - r_0) \diamond (1 - r_3) \leq (1 - r_0) \diamond (1 - r_2) \leq s \leq r.$$

Thus $z \in \mathcal{B}_x^c(r,t)(T,F)$ and hence, we get

$$\mathcal{B}_y^c(1 - r_3, t - t_0)(T,F) \subset \mathcal{B}_x^c(r,t)(T,F).$$

Remark 3.2. $\mathcal{M}^I_{(\mu,\nu)}(T,F)$ is an IFNS.
Define

$$\tau^I_{(\mu,\nu)}(T,F) = A \subset \mathcal{M}^I_{(\mu,\nu)}(T,F) : \text{ for each } x \in A \text{ there exists } t > 0$$
$$\text{and } r \in (0,1) \text{ such that } \mathcal{B}_x(r,t)(T,F) \subset A.$$

Then $\tau^I_{(\mu,\nu)}(T,F)$ is a topology on $\mathcal{M}^I_{(\mu,\nu)}(T,F)$.

In the above result we can easily verify that the open sets in these spaces are open ball in the same spaces. This theorem itself will have various applications in our future work.

Theorem 3.2. The topology $\tau^I_{(\mu,\nu)}(T,F)$ on $\mathcal{M}^I_{0(\mu,\nu)}(T,F)$ is first countable.

Proof. $\{\mathcal{B}_x(\frac{1}{n},\frac{1}{n})(T,F) : n = 1,2,3,...\}$ is a local base at x, the topology $\tau^I_{(\mu,\nu)}(T,F)$ on $\mathcal{M}^I_{0(\mu,\nu)}(T,F)$ is first countable. □

Theorem 3.3. $\mathcal{M}^I_{(\mu,\nu)}(T,F)$ and $\mathcal{M}^I_{0(\mu,\nu)}(T,F)$ are Hausdorff spaces.

Proof. Let $x, y \in \mathcal{M}^I_{(\mu,\nu)}(T,F)$ such that $x \neq y$. Then $0 < F\left(\frac{\mu(T(x)-T(y),t)}{\rho}\right) < 1$ and $0 < F\left(\frac{\nu(T(x)-T(y),t)}{\rho}\right) < 1$.

Putting $r_1 = F\left(\frac{\mu(T(x)-T(y),t)}{\rho}\right)$, $r_2 = F\left(\frac{\nu(T(x)-T(y),t)}{\rho}\right)$ and $r = \max\{r_1, 1-r_2\}$. For each $r_0 \in (r,1)$ there exists r_3 and r_4 such that $r_3 * r_4 \geq r_0$ and $(1-r_3)\diamond(1-r_4) \leq (1-r_0)$.

Putting $r_5 = \max\{r_3, 1-r_4\}$ and consider the open balls $\mathcal{B}_x(1-r_5,\frac{t}{2})$ and $\mathcal{B}_y(1-r_5,\frac{t}{2})$. Then clearly $\mathcal{B}^c_x(1-r_5,\frac{t}{2}) \cap \mathcal{B}^c_y(1-r_5,\frac{t}{2}) = \phi$. For if there exists $z \in \mathcal{B}^c_x(1-r_5,\frac{t}{2}) \cap \mathcal{B}^c_y(1-r_5,\frac{t}{2})$, then

$$r_1 = F\left(\frac{\mu(T(x)-T(y),t)}{\rho}\right) \geq \left(\frac{\mu\left(T(x)-T(z),\frac{t}{2}\right)}{\rho}\right) * F\left(\frac{\mu\left(T(z)-T(y),\frac{t}{2}\right)}{\rho}\right)$$

$$\geq r_5 * r_5 \geq r_3 * r_3 \geq r_0 > r_1$$

and

$$r_2 = F\left(\frac{\nu(T(x)-T(y),t)}{\rho}\right) \leq F\left(\frac{\nu\left(T(x)-T(z),\frac{t}{2}\right)}{\rho}\right) \diamond F\left(\frac{\nu\left(T(z)-T(y),\frac{t}{2}\right)}{\rho}\right)$$

$$\leq (1-r_5)\diamond(1-r_5) \leq (1-r_4)\diamond(1-r_4) \leq (1-r_0) < r_2$$

which is a contradiction. Hence, $\mathcal{M}^I_{(\mu,\nu)}(T,F)$ is Hausdorff. Similarly the proof follows for $\mathcal{M}^I_{0(\mu,\nu)}(T,F)$. □

4. Conclusion

The concept of defining intuitionistic fuzzy ideal convergent sequence spaces as it generalized the fuzzy set theory and give quite useful and interesting applications in many areas of mathematics and engineering. This chapter give brief introduction to intuitionistic fuzzy normed spaces with some basic definitions of convergence applicable on it. We have also summarized different types of sequence spaces with the help of ideal, Orlicz function and compact operator. At the end of this chapter some theorems and remarks based on these new defined sequence spaces are discussed for proper understanding.

Conflict of interest

The authors declare that they have no competing interests.

Author details

Vakeel Ahmad Khan, Hira Fatima and Mobeen Ahmad
Department of Mathematics, Aligarh Muslim University, Aligarh, India

*Address all correspondence to: vakhanmaths@gmail.com

References

[1] Atanassov KT. Intuitionistic fuzzy sets. Fuzzy Sets and Systems. 1986; **20**(1):87-96

[2] Zadeh LA. Fuzzy sets. Information and Control. 1965;**8**:338-353

[3] Barros LC, Bassanezi RC, Tonelli PA. Fuzzy modelling in population dynamics. Ecological Modelling. 2000; **128**:27-33

[4] Hong L, Sun JQ. Bifurcations of fuzzy non-linear dynamical systems. Communications in Nonlinear Science and Numerical Simulation. 2006;**1**:1-12

[5] Fradkov AL, Evans RJ. Control of chaos: Methods of applications in engineering. Chaos, Solitons & Fractals. 2005;**29**:33-56

[6] Giles R. A computer program for fuzzy reasoning. Fuzzy Sets and Systems. 1980;**4**:221-234

[7] Saddati R, Park JH. On the intuitionistic fuzzy topological spaces. Chaos, Solution and Fractals. 2006;**27**: 331-344

[8] Karakus S, Demirci K, Duman O. Statistical convergence on intuitionistic fuzzy normed spaces. Chaos, Solitons and Fractals. 2008;**35**:763-769

[9] Coker D. An introduction to intuitionistic fuzzy topological spaces. Fuzzy Sets and Systems. 1997;**88**(1): 81-89

[10] Mursaleen M, Lohani QMD. Intuitionistic fuzzy 2-normed space and some related concepts. Chaos, Solution and Fractals. 2009;**42**:331-344

[11] Khan VA, Ebadullah K, Rababah RKA. Intuitionistic fuzzy zweier I-convergent sequence spaces. Functional Analysis: Theory, Methods and Applications. 2015;**1**:1-7

[12] Steinhaus H. Sur la convergence ordinaire et la convergence asymptotique. Colloquium Mathematicum. 1951;**2**:73-74

[13] Fast H. Sur la convergence statistique. Colloquium Mathematicum. 1951;**2**:241-244

[14] Schoenberg IJ. The integrability of certain functions and related summability methods. American Mathematical Monthly. 1959;**66**:361-375

[15] Kostyrko P, Salat T, Wilczynski W. *I*-convergence. Real Analysis Exchange. 2000;**26**(2):669-686

[16] Alotaibi A, Hazarika B, Mohiuddine SA. On the ideal convergence of double sequences in locally solid Riesz spaces. Abstract and Applied Analysis. 2014. 6 p. Article ID: 396254. http://dx.doi.org/10.1155/2014/396254

[17] Hazarika B, Mohiuddine SA. Ideal convergence of random variables. Journal of Function Spaces and Applications. 2013. Article ID 148249:7

[18] Mohiuddine SA, Lohani QMD. On generalized statistical convergence in intuitionistic fuzzy normed spaces. Chaos, Solitons and Fractals. 2009;**41**: 142-149

[19] Mursaleen M, Mohiuddine SA, Edely OHH. On the ideal convergence of double sequences in intuitionistic fuzzy normed spaces. Computers and Mathematics with Application. 2010;**59**: 603-611

[20] Nabiev A, Pehlivan S, Gürdal M. On *I*-Cauchy sequence. Taiwanese Journal of Mathematics. 2007;**11**(2):569-576

[21] Bromwich TJI. An Introduction to the Theory of Infinite Series. New York: MacMillan Co. Ltd; 1965

[22] Khan VA, Fatima H, Abdullaha SAA, Khan MD. On a new BV_σ I-convergent double sequence spaces. Theory and Application of Mathematics and Computer Science. 2016;**6**(2): 187-197

[23] Khan VA, Shafiq M, Guillen BL. On paranorm I-convergent sequence spaces defined by a compact operator. Afrika Matematika, Journal of the African Mathematical Union (Springer). 2014; **25**(4):12. DOI: 10.1007/s13370-014-0287-2

[24] Kreyszig E. Introductory Functional Analysis with Application. New York, Chicheste, Brisbane, Toronto: John Wiley and Sons, Inc; 1978

Section 3

Adaptive Neuro-Fuzzy Inference Systems

Chapter 3

ANFIS: Establishing and Applying to Managing Online Damage

Sy Dzung Nguyen

Abstract

Fuzzy logic (FL) and artificial neural networks (ANNs) own individual advantages and disadvantages. Adaptive neuro-fuzzy inference system (ANFIS), a fuzzy system deployed on the structure of ANN, by which FL and ANN can interact to not only overcome their limitations but also promote the ability of each model has been considered as a reasonable option in the real fields. With the vital strong points, ANFIS has been employed well in many technology applications related to filtering, identifying, predicting, and controlling noise. This chapter, however, focuses mainly on building ANFIS and its application to identifying the online bearing fault. First, a traditional structure of ANFIS as a data-driven model is shown. Then, a recurrent mechanism depicting the relation between the processes of filtering impulse noise (IN) and establishing ANFIS from a noisy measuring database is presented. Finally, one of the typical applications of ANFIS related to online managing bearing fault is shown.

Keywords: fuzzy logic, artificial neural networks, adaptive neuro-fuzzy inference system

1. Introduction

As well known, the mathematical tools FL and ANN possess both the advantages and disadvantages as their specific characteristics. The hybrid structure ANFIS, where ANN and FL can interact to not only overcome partly the limitations of each model but also uphold their strong points [1–25], is, hence, considered as a reasonable option in many technology applications such as identifying [1–2, 4, 6, 12], predicting [9, 11, 17, 25], controlling [3, 5, 7, 18–26], and filtering noise [14–16, 27–29].

To build an ANFIS from a given database, firstly, an initial data space (IDS) expressing the mapping $f: X \rightarrow Y$ must be created. A cluster data space (CDS) is then built from the IDS to form the ANFIS via a training algorithm. Being viewed as a popular technique for unsupervised pattern recognition, clustering is an effective tool for analyzing and exploring data structures to build CDSs [30–37]. Reality shows that the accuracy and training time of the ANFIS depend deeply on the features of both the IDS and CDS [2–6]. In the process of building ANFIS, the two issues as follows should be considered: (1) What is the essence of the interactive relation between ANFIS's convergence capability and CDS' attributes? (2) How to exploit this essence for increasing ANFIS's ability to converge to the desired accuracy with the improved calculating cost?

Many different clustering approaches have been discovered [2–4, 10, 30–34, 37]. Separating data in X and in Y distinctly with a mutual result reference, step by step, was described in [10]. The method, however, could not solve appropriately the above issues. Besides, the difficulty in deploying fuzzy clustering strategies along with the high calculating cost was their disadvantage. Generally, a hard relation could not reflect fully database attributes [31, 34]. The well-known method of fuzzy C-means clustering was seen as a better option in this case. It, however, was not effective enough for the “non-spherical” general datasets [30, 37]. Therefore, the idea of fuzzy clustering in a kernel feature space was then developed to deal with these cases [30–34, 37]. A weighted kernel-clustering algorithm could be referred to [30], or a method of weighted kernel fuzzy C-means clustering based on adaptive distances was detailed in [31]. In spite of owning considerable advantages, the identification and prediction accuracy of the ANFIS based on the CDS coming from [30–31] are sensitive to attributes of the CDS due to the negative influence of noise.

Reality has shown that noise status including IN always exists in the measured IDSs [2, 4, 9, 16], which degrades violently the accuracy of ANFISs deriving from them. There are many reasons resulting in this, such as the lack of precision of the measurement devices, tools, measurement methods, or the negative impact of the surrounding environment. In [7], an ANFIS took part in the system in the form of an inverse MR damper (MRD) model to specify the time-verifying desired control current. To maintain the accuracy of the inverse MRD model, the ANFIS was retrained after each certain period due to the dynamic response of the MRD depending quite deeply on temperature. Another more active approach is filtering noise or preprocessing data [7, 9, 11, 17, 21, 38–40]. In [11, 17], where ANFISs were employed to predict the health of mechanical systems, vibration signal was always measured and filtered to update the ANFISs. Related to the preprocessing data to set up ANFIS, it can observe that to maintain the stability of the above online ANFIS-based applications, reducing time delay is really meaningful. One of the becoming solutions for this can be referred to [16] where filtering IN and building ANFIS were carried out synchronously via a recurrent mechanism. A recurrent strategy for forming ANFIS was carried out, in which the capability to converge to a desired accuracy of the ANFIS training process could be estimated and directed online. As a solution, increasing the quality of both the IDS and CDS was paid attention. Building an ANFIS via a filtered database and exploiting the ANFIS as an updated filter to refilter the database were depicted via an online and recurrent mechanism. The process was upholden until the ANFIS-based database approximation convergent to the desired accuracy or a stop condition appears.

Inspired by the ANFIS's capability, in order to provide the readers with the theoretical basis and application direction of the model, this chapter presents the formulation of ANFIS and one of its typical applications. The rest of the chapter is organized as follows. Section 2 shows a structure of ANFIS as a data-driven model deriving from fuzzy logic and artificial neural networks. Setting up the CDS consisting of the input data clusters, output data clusters, and the CDS-based ANFIS as a jointed structure is all detailed. Deriving from this relation, a theoretical basis for building ANFIS from noisy measuring datasets is presented in Section 3. An online and recurrent mechanism for filtering noise and building ANFIS synchronously is clarified via algorithms for filtering IN and establishing ANFIS. A typical application of ANFIS related to online managing bearing fault status is shown in Section 4. Finally, some general aspects are mentioned in the last section.

2. Structure of ANFIS

Let's consider a given IDS having P input–output data samples $(\overline{x}_i, y_i)$, $\overline{x}_i = [x_{i1}, \dots, x_{in}] \in \Re^n$, $y_i \in \Re^1$, and $i = 1 \dots P$. With a data normalization solution and a used certain clustering algorithm, a CDS is then created. The kth cluster, signed Γ^k, $k = 1...C$, consists of one input cluster and one output cluster signed $\Gamma^{k(A)}$ and $\Gamma^{k(B)}$, respectively. The CDS can be seen as a framework for establishing ANFIS. This section presents how to build the CDS as well as the CDS-based ANFIS structure.

2.1 Some related notions

Some notions shown in [16] are used in this chapter as follows.

Definition 1. Normalizing a given IDS to set up a normalized initial data space signed $\overline{\text{IDS}}$ is performed as follows:

$$\tilde{x}_{ij} = x_{ij} / \max_{k} |x_{kj}|, \qquad i,\ k = 1...P, j = 1...n. \tag{1}$$

By this way, the ith data sample (also signed $(\overline{x}_i, y_i)$) in the $\overline{\text{IDS}}$ is constituted as follows:

$$\left(\overline{x}_i = \left[\tilde{x}_{i1}, ..., \tilde{x}_{in}\right]^T, y_i\right) \quad i = 1...P. \tag{2}$$

Definition 2. The root-mean-square error (RMSE) in Eq. (3) is used to evaluate accuracy rate of ANFIS. The required RMSE value is signed $[E]$. The absolute error, $\overline{\varepsilon}_i, i = 1...P,$ between the data output $y_i = f(\overline{x}_i)$ and the corresponding ANFIS-based output $\hat{y}_i(\overline{x}_i)$ is defined in Eq. (4). The desired value of $\overline{\varepsilon}_i$ is signed $[\overline{\varepsilon}]$:

$$\text{RMSE} = \sqrt{P^{-1}\textstyle\sum_{i=1}^{P}\left(\hat{y}_i(\overline{x}_i) - f(\overline{x}_i)\right)^2}, \tag{3}$$

$$\overline{\varepsilon}_i = \left|\hat{y}_i(\overline{x}_i) - f(\overline{x}_i)\right| \ , i = 1...P. \tag{4}$$

Definition 3. Let's consider $\overline{x}_i \in \overline{\text{IDS}}$ in which $\overline{\text{IDS}}$ depicts an unknown mapping $f : \text{X} \to \text{Y}$. The ANFIS-based approximation of $f : \text{X} \to \text{Y}$ is called to be continuous at $\left(\overline{x}_p, y_p\right) \in \overline{\text{IDS}}$ if.

$$\hat{y}_i(\overline{x}_i) \to y_p \pm [\overline{\varepsilon}] \text{ when } \overline{x}_i \to \overline{x}_p. \tag{5}$$

Definition 4. Let's consider an ANFIS-based approximation of a mapping expressed by an $\overline{\text{IDS}}$. The ANFIS is said to be a uniform approximation with a required error $[\overline{\varepsilon}]$ if at $\forall\, \overline{x}_i \in \text{X}$, by choosing any small constant $\varepsilon \geq [\overline{\varepsilon}]$, corresponding data sample $\overline{x}_j \in \overline{\text{IDS}}$ always exists such that.

$$\forall\, \overline{x}_h \in \overline{\text{IDS}}; \text{if } \|\overline{x}_h - \overline{x}_i\| \leq \left\|\overline{x}_j - \overline{x}_i\right\|, \text{then } \left|\hat{y}_h(\overline{x}_h) - f(\overline{x}_i)\right| \leq \left|\hat{y}_j(\overline{x}_j) - f(\overline{x}_i)\right| = \varepsilon. \tag{6}$$

Definition 5. Data cluster Γ^k and data sample $\left(\overline{x}_p, y_p\right) \in \Gamma^k$ in a CDS derived from an $\overline{\text{IDS}}$ are depicted in **Figure 1**. Let sign $\Gamma^k \backslash p$ to be a subset consisting of the data samples belonging to Γ^k except $\left(\overline{x}_p, y_p\right)$. The subset contains Q_{kp} data samples. It is assumed that all of data samples in $\Gamma^{k(A)}$ are distributed closely, while in $\Gamma^{k(B)}$, most of them are located closely, except y_p; it is far from the other and distributes at one side of $\Gamma^{k(B)}$. This status is described in Eq. (7):

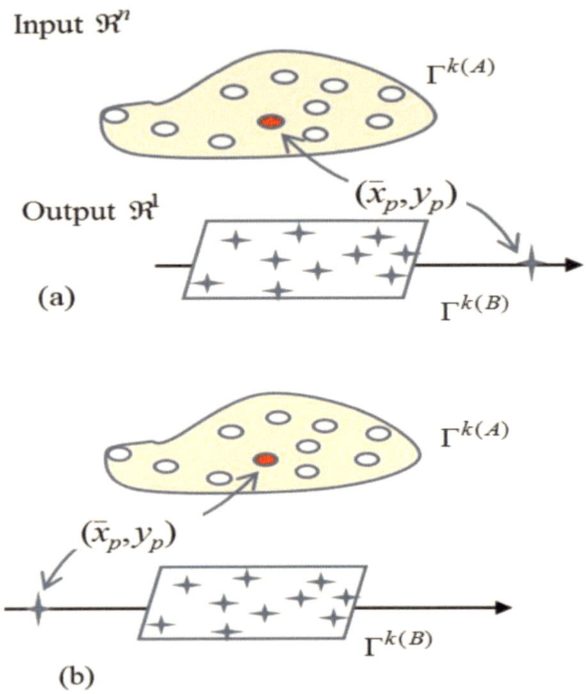

Figure 1.
Two typical distribution types in data cluster Γ^k: Impulse noise point IN $\left(\bar{x}_p, y_p\right) \in \Gamma^k$ causing the distribution at one side, the right side (a), and the left side (b).

$$y_p \gg \max_{y_i \in \Gamma^{k(B)} \backslash p} (y_i) \text{ or } y_p \ll \min_{y_i \in \Gamma^{k(B)} \backslash p} (y_i). \tag{7}$$

and satisfies Eqs. (8) and (9):

$$d_{k1} = y_p - \max_{y_i \in \Gamma^{k(B)} \backslash p} (y_i) > \left(P[E]^2 - Q_{kp}[\bar{\varepsilon}]^2\right)^{0.5} \quad \text{if} \quad y_p > \max_{y_i \in \Gamma^{k(B)} \backslash p} (y_i). \tag{8}$$

$$d_{k2} = \min_{y_i \in \Gamma^{k(B)} \backslash p} (y_i) - y_p > \left(P[E]^2 - Q_{kp}[\bar{\varepsilon}]^2\right)^{0.5} \quad \text{if} \quad y_p < \min_{y_i \in \Gamma^{k(B)} \backslash p} (y_i). \tag{9}$$

In this case, $\left(\bar{x}_p, y_p\right)$ is called a *critical data sample* in the CDS.

2.2 Setting up the input data clusters

Let's consider the normalized initial data space $\overline{\text{IDS}}$ (see Def. 1). Many well-known clustering methods can be used to build a CDS from the $\overline{\text{IDS}}$. Here, the CDS is built by using the clustering algorithm KFCM-K (kernel fuzzy C-means clustering with kernelization of the metric) presented in [31]. By this way, distribution of data samples in the CDS is established. The membership degree of the jth data sample belonging to the ith cluster is denoted by $\mu_{ij} \in [0, 1]\ \forall i, j$ and $j = 1...P, i = 1...C$. Cluster centroids $\bar{x}_1^0, ..., \bar{x}_C^0$ in the CDS are specified such that the following objective function is minimized:

$$J_{KFCM}(U,\overline{x}^0) = \sum_{i=1}^{C}\sum_{j=1}^{P} \mu_{ij}{}^m \|\phi(\overline{x}_j) - \phi(\overline{x}_i^0)\|^2 \tag{10}$$

subjected to $\sum_{i=1}^{C} \mu_{ij} = 1\,\forall j$ and $\mu_{ij} \in [0,1]\ \forall i,j$. In Eq. (10), $\overline{x}_i^0 = [x^0{}_{i1}, ..., x^0{}_{in}] \in \Re^n$ is the ith cluster center; $\|\phi(\overline{x}_j) - \phi(\overline{x}_i^0)\|^2$ denotes the squared distance between $\overline{x}_j$ and $\overline{x}_i^0$ in the kernel space; $\phi(.)$ is the kernel function; $U = U(\mu_{ij}) \in \Re^{(C\times P)}$ is the distribution matrix; and $m > 1$ is the fuzzy factor.

The objective function can be rewritten via Gaussian kernel function as follows:

$$J_{KFCM}(U,\overline{x}^0) = 2\sum_{i=1}^{C}\sum_{j=1}^{P} \mu_{ij}{}^m \left(1 - \exp\left(-\|\overline{x}_j - \overline{x}_i^0\|^2/\sigma^2\right)\right). \tag{11}$$

Deriving $J_{KFCM}(U,\overline{x}^0)$ in Eq. (11) with respect to $\overline{x}_i^0$, at the optimal centers, the following must be obtained:

$$\frac{\partial}{\partial \overline{x}_i^0} J_{KFCM}(U,\overline{x}_i^0) = \frac{-4}{\sigma^2}\sum_{j=1}^{P} \mu_{ij}{}^m \ (\overline{x}_j - \overline{x}_i^0) \exp\left(-\|\overline{x}_j - \overline{x}_i^0\|^2/\sigma^2\right) = 0 \tag{12}$$

From Eqs. (11) to (12) and the use of Lagrange multipliers with $\mu_{ij} \in [0,1]\ \forall i,j$ and $\sum_{i=1}^{C}\mu_{ij} = 1\forall j$, the following update laws are obtained:

$$\overline{x}_i^0 = \frac{\sum_{j=1}^{P} \mu_{ij}{}^m \ \ \overline{x}_j \ K(\overline{x}_j, \overline{x}_i^0)}{\sum_{j=1}^{P} \mu_{ij}{}^m K(\overline{x}_j, \overline{x}_i^0)}, i = 1...C. \tag{13}$$

$$\mu_{ij} = \begin{cases} \left[\left(\sum_{h=1}^{C} \frac{1 - K(\overline{x}_j, \overline{x}_i^0)}{1 - K(\overline{x}_j, \overline{x}_h^0)}\right)^{1/(m-1)}\right]^{-1} & \text{if } \overline{x}_j \neq \overline{x}_i^0 \\ 1 \ \left(\text{and } \mu_{ik(k\neq j)} = 0\right) \ \text{ if } \overline{x}_j = \overline{x}_i^0 & \end{cases} \tag{14}$$
$$i = 1...C; j = 1...P.$$

By using index ts as in Eq. (15), $[ts]$ to be the required value of ts and r to denote the rth loop, the clustering phase is accomplished until $ts \leq [ts]$:

$$ts = \left(J_{KFCM}{}^{(r)} - J_{KFCM}{}^{(r-1)}\right)/J_{KFCM}{}^{(r-1)} \tag{15}$$

Specification of the optimal centers and their relationship values as abovementioned is detailed in Appendix A of [12].

2.3 Setting up the output data clusters

The result of the clustering process in the input data space is an input cluster centroid vector $[\overline{x}_1^0, ..., \overline{x}_C^0]$ of corresponding data clusters, respectively, signed as $\Gamma^1, ..., \Gamma^C$. Let $A^1, ..., \ A^C$, respectively, be input fuzzy sets established via $\overline{x}_1^0, ..., \overline{x}_C^0$ [12, 16]. The membership value of $\tilde{x}_{il}$ belonging to A^k is inferred from Eq. (14):

$$\bar{\mu}_{ki}\left(\tilde{x}_{il}\right) = \left[\left(\sum_{h=1}^{C}\frac{1-K\left(\tilde{x}_{il},x_{kl}^{0}\right)}{1-K\left(\tilde{x}_{il},x_{hl}^{0}\right)}\right)^{1/(m-1)}\right]^{-1}. \tag{16}$$
$$(k = 1...C; i = 1...P;\ l = 1...n.)$$

With following the product law, membership value of $\bar{x}_q$ belonging to A^i is

$$\mu_{kq}\left(\bar{x}_q\right) = \prod_{l=1}^{n}\bar{\mu}_{kq}\left(\tilde{x}_{ql}\right),\quad k = 1...C,\ q = 1...P, \tag{17}$$

and its normalized membership value is as follows:

$$N_k\left(\bar{x}_q\right) = \mu_{kq}\left(\bar{x}_q\right)/\sum_{h=1}^{C}\mu_{hq}\left(\bar{x}_q\right),\quad q = 1...P, k = 1...C. \tag{18}$$

The membership of a data sample in each cluster determined based on Eqs. (16)–(18) is then used to specify the hard distribution status of the data samples in each cluster. It is then used to specify the index vector **a** of hyperplanes (or the output data clusters) $w_k(.)$ and $k = 1...C$. The ith data sample is hardly distributed into the kth data cluster if

$$N_k(\bar{x}_i) = \max_{h=1...C}\left(N_h(\bar{x}_i)\right),\quad i = 1...P,\ k = 1...C. \tag{19}$$

Deriving from the t_k data samples hardly distributed in the kth data cluster, by using the least mean squares method, vector $\mathbf{a} = [a_0, a_1, ..., a_n]^T = [a_0, \bar{\mathbf{a}}]^T$ of $w_k(.)$ is specified which is the solution of Eq. (20):

$$\begin{cases} a_n\sum_{i=1}^{t_k}\tilde{x}_{in} + a_{n-1}\sum_{i=1}^{t_k}\tilde{x}_{i(n-1)} + ... + a_1\sum_{i=1}^{t_k}\tilde{x}_{i1} + a_0 t_k = \sum_{i=1}^{t_k}y_i \\ a_n\sum_{i=1}^{t_k}\tilde{x}_{in}\tilde{x}_{i1} + a_{n-1}\sum_{i=1}^{t_k}\tilde{x}_{i(n-1)}\tilde{x}_{i1} + ... + a_1\sum_{i=1}^{t_k}\tilde{x}_{i1}^{\ 2} + a_0\sum_{i=1}^{t_k}\tilde{x}_{i1} = \sum_{i=1}^{t_k}y_i\tilde{x}_{i1} \\ \vdots \\ a_n\sum_{i=1}^{t_k}\tilde{x}_{in}^{\ 2} + a_{n-1}\sum_{i=1}^{t_k}\tilde{x}_{i(n-1)}\tilde{x}_{in} + ... + a_1\sum_{i=1}^{t_k}\tilde{x}_{i1}\tilde{x}_{in} + a_0\sum_{i=1}^{t_k}\tilde{x}_{in} = \sum_{i=1}^{t_k}y_i\tilde{x}_{in}. \end{cases} \tag{20}$$

Finally, the value of hyperplane w_k corresponding to $\bar{x}_i$ is calculated in Eq. (21):

$$w_k(\bar{x}_i) = a_0 + \bar{\mathbf{a}}^T\bar{x}_i \tag{21}$$

2.4 Structure of ANFIS

As mentioned in Eq. (1), the ANFIS for approximating the mapping $f : \mathrm{X} \to \mathrm{Y}$ is derived from M fuzzy laws in Eq. (22):

$$R^{(i)} : \text{IF } \tilde{x}_{q1} \text{ is } A_1^i\left(\tilde{x}_{q1}\right), \text{AND...}, \text{AND } \tilde{x}_{qn} \text{ is } A_n^i\left(\tilde{x}_{qn}\right) \text{THEN } y_q^i \text{is } B^i\left(\bar{x}_q\right) \tag{22}$$
$$(i = 1...\mathrm{M}, \mathrm{M} \equiv C),$$

where $A_l^i\left(\tilde{x}_{ql}\right), i = 1...\mathrm{M}, l = 1...n,$ is the membership value of $\tilde{x}_{ql}$ belonging to input fuzzy set A^i, meaning $A_l^i(\bar{x}_q) \equiv \bar{\mu}_{iq}\left(\tilde{x}_{ql}\right)$ in Eq. (16); $B^i\left(\bar{x}_q\right)$ is the corresponding output fuzzy set of data sample $\bar{x}_q, \bar{x}_q = \left[\tilde{x}_{q1}, ..., \tilde{x}_{qn}\right]^T,\ q = 1...P.$

In the fuzzification phase, membership value of $\overline{x}_q$ belonging to input fuzzy set A^i signed $A^i(\overline{x}_q) \equiv \mu_{iq}(\overline{x}_q)$ is specified by Eq. (17). For the defuzzification, if the center-average method is used, the output of the qth data sample is expressed via the membership values in the input fuzzy space of $\overline{x}_q$ as follows:

$$\hat{y}_q(\overline{x}_q) = \sum_{i=1}^{M} y_q^i \; \mu_{iq}(\overline{x}_q) / \sum_{i=1}^{M} \mu_{iq}(\overline{x}_q) \tag{23}$$

where $y_q^i = w_i(\overline{x}_q)^i$ is the value of hyperplane w_i corresponding to data sample $\overline{x}_q$ calculated in Eq. (21).

Finally, all the above-mentioned contents can be depicted via the ANFIS with five layers signed D, CL, Π, N, and S in **Figure 2**. Layer D (data) has n input nodes corresponding to n elements of data vector $\overline{x}_i = [x_{i1}, \ldots, x_{in}]^T$, i = 1...P, while its outputs are the corresponding normalized values using Eq. (1). Layer CL (clustering) expresses the clustering process. The result of this process is C clusters with C corresponding cluster centroids $\overline{x}_1^0, \ldots, \; \overline{x}_C^0$, to which C fuzzy sets, $A^1, \ldots, A^C$, are given. The output of this layer is the membership value of $\overline{x}_i$ calculated for each dimension $(\tilde{x}_{i1}, \ldots, \tilde{x}_{in})$ via Eq. (16). Layer Π (product layer) specifies membership values based on Eq. (17). Layer N (normalization) estimates the normalized membership value of a data sample belonging to each fuzzy set upon Eq. (18). Layer S (specifying) is used to estimate the output of the ANFIS based on any well-known method. In case of using the center-average defuzzification, it is calculated by Eq. (23), while it is specified by Eq. (24) if the "the winner takes all" law is employed:

$$\hat{y}_i = w_k(\overline{x}_i), \quad i = 1 \ldots P, \tag{24}$$

where $w_k(\overline{x}_i)$ is the value of the kth hyperplane corresponding to input data sample $\overline{x}_i$ (21); k is the index of the data cluster where data sample $\overline{x}_i$ gets the maximum membership specified via $N_{(.)}(\overline{x}_i)$ as in Eq. (25):

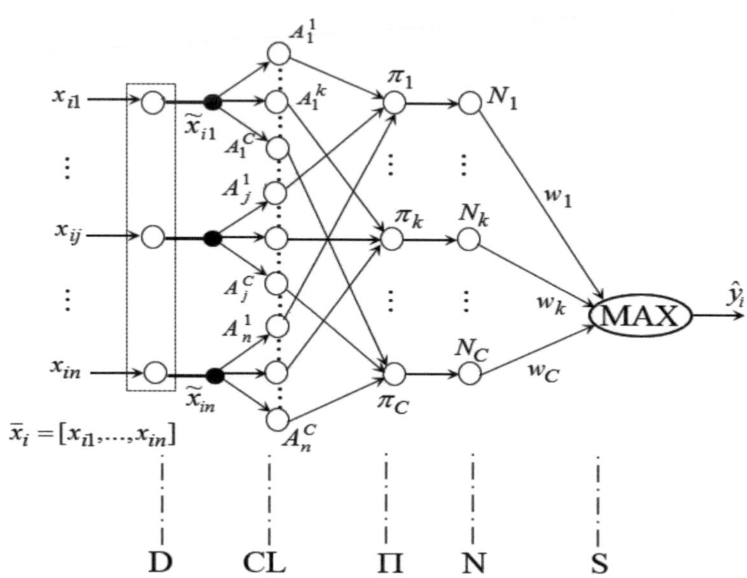

Figure 2.
Structure of the ANFIS.

$$N_k(\overline{x}_i) = \max_{h=1...C} (N_h(\overline{x}_i)) \tag{25}$$

3. Building ANFIS from a noise measuring database

This section presents the recurrent mechanism together with the related algorithms consisting of the one for ANFIS-based noise filtering and the one for building ANFIS showed in [16].

3.1 Convergence condition of the ANFIS-based approximation

Deriving from a given IDS having P input–output data samples $(\overline{x}_i, y_i)$, $\overline{x}_i = [x_{i1}, ..., x_{in}] \in \mathfrak{R}^n$ and $y_i \in \mathfrak{R}^1$, $i = 1...P$, with a data normalization solution as in Def. 1, the $\overline{\text{IDS}}$ is built, to which a CDS is created as depicted in Section 2. It should be noted that IN is often considered as disturbances distributed uniformly in a signal source which impacts negatively on the created CDS. In general, IN causes raising the number of critical data samples in the CDS. The negative impact of IN on the convergent ability of training ANFIS is formulated via Theorem 1 as follows.

Theorem 1 [16]: Let's consider a given $\overline{\text{IDS}}$ deriving from an IDS and an ANFIS uniformly approximating an unknown mapping $f : \text{X} \to \text{Y}$ expressed by the IDS. The ANFIS is built via a CDS built from the $\overline{\text{IDS}}$. Assume that X is compact. The necessary condition for the approximation convergent to a desired error $[E]$ is that in the CDS there is not any critical data sample.

Proof: Let's consider cluster Γ^k belonging to the CDS. Assume that $\left(\overline{x}_p, y_p\right) \in \Gamma^k$ is a critical data sample (see Def. 5); it has to be proven that the ANFIS will not converge to $[E]$.

It can infer from Eq. (3) that

$$\text{RMSE} \geq P^{-0.5}\left(\left(\hat{y}_p(\overline{x}_p) - f(\overline{x}_p)\right)^2 + \sum_{i=1}^{Q_{kp}} (\hat{y}_i(\overline{x}_i) - f(\overline{x}_i))^2\right)^{0.5} \tag{26}$$

Because the ANFIS is a uniform approximation of $f : \text{X} \to \text{Y}$ and X is the compact set, it can infer that the ANFIS is continuous in $\Gamma^k \backslash p$, so Eq. (27) can be inferred from Eq. (26):

$$\text{RMSE} \geq P^{-0.5}\left(\left(\hat{y}_p(\overline{x}_p) - f(\overline{x}_p)\right)^2 + Q_{kp}[\overline{\varepsilon}]^2\right)^{0.5} \tag{27}$$

It should be noted that the ANFIS is a uniform approximation of the $f : \text{X} \to \text{Y}$ in $\Gamma^k \backslash p$, $\left(\overline{x}_p, y_p\right) \in \Gamma^k$ is a critical data sample, and samples in $\Gamma^{k(A)}$ are distributed closely. As a result, Eq. (28) can be inferred:

$$\hat{y}_p(\overline{x}_p) \in \left[\min_{y_i \in \left(\Gamma^{k(B)} \backslash p\right)} (y_i) \quad \max_{y_i \in \left(\Gamma^{k(B)} \backslash p\right)} (y_i)\right] \tag{28}$$

Due to $y_p > \max_{y_i \in \left(\Gamma^{k(B)} \backslash p\right)} (y_i)$ (see Def. 5), the following can be obtained from Eqs. (27) to (28):

$$\text{RMSE} \geq P^{-0.5}\left(\left(\max_{y_i \in (\Gamma^{k(B)} \backslash p)}(y_i) - y_p\right)^2 + Q_{kp}[\overline{\varepsilon}]^2\right)^{0.5} \tag{29}$$

From Eqs. (8) and (29), it can conclude that RMSE > $[E]$. Similarly, due to $y_p < \min_{y_i \in (\Gamma^{k(B)} \backslash p)}(y_i)$ (see Def. 5), from Eqs. (27) to (28), the following can be also inferred:

$$\text{RMSE} \geq P^{-0.5}\left(\left(\min_{y_i \in (\Gamma^{k(B)} \backslash p)}(y_i) - y_p\right)^2 + Q_{kp}[\overline{\varepsilon}]^2\right)^{0.5} \tag{30}$$

From Eq. (9) to (30), RMSE > $[E]$ can be implied.

Finally, it can conclude that if existing at least a critical data sample in the CDS, the ANFIS could not converge to the required error $[E]$. □.

3.2 Algorithm for filtering IN

An essential advantage of the clustering algorithms presented in [30–31] is the convergent rate. However, the quality of the ANFIS based on the CDS deriving from them is sensitive to the IDS attributes. It can be observed that the main reason

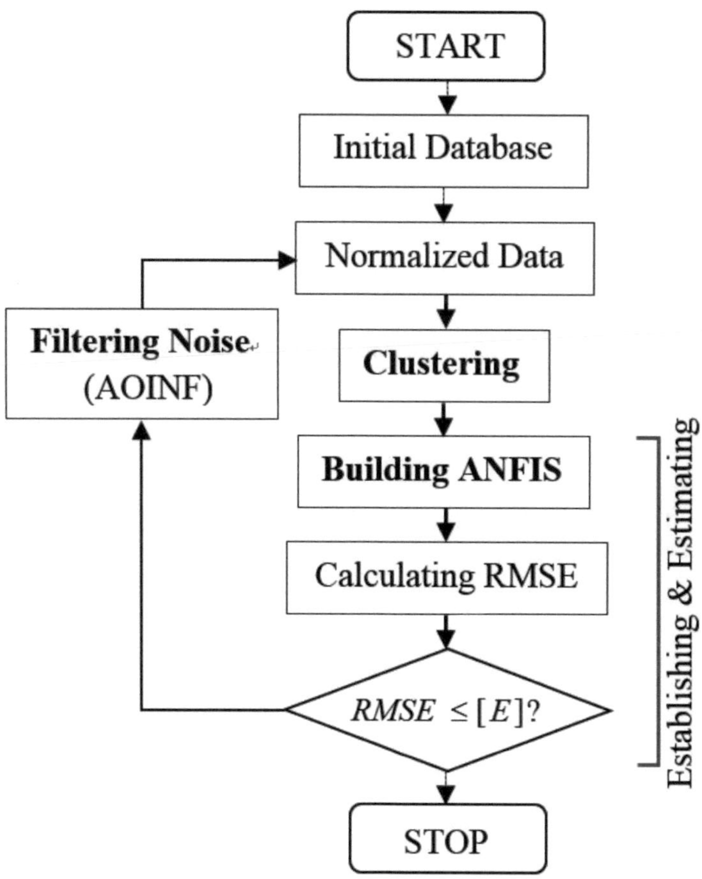

Figure 3.
Flowchart of the FIN-ANFIS consisting of the three main phases, the clustering, establishing and estimating ANFIS, and filtering IN, which are performed simultaneously.

of this status via Theorem 1 is the appearance of critical data samples. Besides, regarding the preprocessing IDS shown in [9], in spite of the positive filtering effectiveness, the calculating cost of the method is quite high. A becoming solution for the above issues can be referred in [16] where the recurrent mechanism illustrated in **Figure 3** was employed. The recurrent mechanism has two phases being performed synchronously: filtering IN in the database and building ANFIS based on the filtered database.

Firstly, an adaptive online impulse noise filter (AOINF) is proposed. The recurrent mechanism is then depicted via the algorithm named FIN-ANFIS consisting of three main phases: filtering IN, clustering data, and building ANFIS. By this way, the filtered $\overline{\text{IDS}}$ is used to build the ANFIS, then the created ANFIS is applied as an updated filter to refilter the $\overline{\text{IDS}}$, and so on, until either the process converges or a stop condition is satisfied. To get a guarantee of convergence and stability, an update law for the AOINF is discovered via Lyapunov stability theory.

Remark 1. ANFIS cannot converge to the required error $[E]$ if there is at least one critical data sample in the CDS (see Theorem 1). The clustering strategy of the FIN-ANFIS therefore focuses on preventing the clustering process from appearing critical data samples, along with seeking to exterminate the critical data samples in the CDS having been taking form. As a result, in each loop of the ANFIS training process, the strategy well directs the clustering process to a new CDS where either there is not any critical data sample or there exist with a smaller amount. Theorem 2 shows the convergence condition of the training process.

Theorem 2 [16]: Following the flowchart in **Figure 3**, the ANFIS-based approximation of an unknown mapping $f: \mathrm{X} \to \mathrm{Y}$ expressed by the given IDS is built via a CDS which drives from the $\overline{\text{IDS}}$ (the normalized IDS). Let Q be the number of critical data points in the CDS at the rth loop. At these critical data samples, if the data output is filtered by law (31), then the RMSE (3) of the ANFIS will converge to $[E]$:

$$^{(r+1)}y_i = {}^{(r)}y_i - \rho \ \text{sgn} \ \left(^{(r)}(y_i - \hat{y}_i) \right), \quad i = 1...Q. \tag{31}$$

In the above, $\rho > 0$ is the update coefficient to be optimized by any well-known optimal method; $^{(r)}(y_i - \hat{y}_i)$ is the error between the ith data output and the corresponding ANFIS-based output; and function $\text{sgn}\,(.)$ is defined as

$$\text{sgn}\,(z) = \begin{cases} 1 & if \quad z > 0 \\ -1 & \text{otherwise.} \end{cases} \tag{32}$$

Proof: A Lyapunov candidate function is chosen as in Eq. (33), to which expression (34) can be inferred:

$$e(\mathbf{X}) = \mathbf{X}^T\mathbf{X}. \tag{33}$$

$$\dot{e}(\mathbf{X}) = 2\sum_{i=1}^{P-Q} X_i\dot{X}_i + 2\sum_{j=1}^{Q} X_j\dot{X}_j. \tag{34}$$

In the above, $\dot{\Xi} = d\Xi/dt$ expresses derivative of Ξ with respect to time; $\mathbf{X}$ is the vector of state variables deriving from $\overline{\text{IDS}}$ as follows:

$$X_i = y_i - \hat{y}_i; \mathbf{X} = [X_1, ..., X_P]^T \tag{35}$$

From update law (31), Eq. (34) can be rewritten as in Eq. (36):

$$\begin{aligned}\dot{e}(\mathbf{X}) &= 2\sum_{i=1}^{P-Q} X_i\dot{X}_i + 2\sum_{j=1}^{Q} X_j\ \dot{y}_j \\ &= 2\sum_{i=1}^{P-Q} X_i\dot{X}_i - 2\rho\sum_{j=1}^{Q} X_j\ \mathrm{sgn}\left(X_j\right).\end{aligned} \tag{36}$$

It should be noted that the update process is performed with respect to the critical data points; hence, Eq. (36) can be rewritten as follows:

$$\dot{e}(\mathbf{X}) = -2\rho\sum_{j=1}^{Q} X_j\ \ \mathrm{sgn}\left(X_j\right) = -2\rho\sum_{j=1}^{Q}\left|X_j\right| < 0 \tag{37}$$

In addition, the following can be implied from (33) to (35):

$$e(0) = 0;\quad e(\mathbf{X}) \geq 0\ \ \forall\mathbf{X}. \tag{38}$$

Finally, it can infer from Eqs. (37) to (38) that $e(\mathbf{X}) \to 0$ is a stable Lyapunov process. Hence, from Eq. (3) one can infer the aspect needing to be proven:

$$\mathrm{RMSE} = \lim_{r\to\infty}\sqrt{{}^{(r)}e(\mathbf{X})P^{-1}}\ \ \leq [E].\ \ \square \tag{39}$$

Remark 2. (1) To enhance the ability to adapt to the noise status of the $\overline{\text{IDS}}$, ρ in Eq. (31) is specified as follows:

$$\rho = \alpha\left|{}^{(r)}\left(y_i - \hat{y}_i\right)\right|, \tag{40}$$

where $\alpha \geq 0$ is an adaptive coefficient chosen by the designer. Thus, $\rho = \rho(X_i, t)$ takes part in adjusting the filtering level $\Delta_i = \left|{}^{(r+1)}y_i - {}^{(r)}y_i\right|$. (2) It can infer from Theorem 1 that disposing of critical data samples in the CDS needs to be carried out. Therefore, the useful solution offered in Theorem 2 via update law (31) is employed to establish the filtering mechanism of the AOINF as shown below.

The algorithm AOINF for filtering IN:

1. Look for critical data samples in the CDS to specify the worst data point (WP) where the continuous status of the ANFIS is worst:

$$\mathrm{WP} \equiv \left(\overline{x}_i^{(WP)}, y_i^{(WP)}\right)\quad \text{such that}\quad \left|y_i^{(WP)} - \hat{y}_i^{(WP)}\right| = \max_{h=1\dots P}\left|y_h - \hat{y}_h\right|. \tag{41}$$

1. Specify the data samples satisfying condition (42):

$$\left|y_q - \hat{y}_q\right| \ \geq\ \frac{1}{\sigma}\left|y_i^{(WP)} - \hat{y}_i^{(WP)}\right|\ , q = 1\dots\ \overline{Q}. \tag{42}$$

In the above, $\hat{y}_i^{(WP)}$ is the ANFIS-based output, while $y_i^{(WP)}$ is the corresponding data output at the WP; $\sigma > 1$ is an adaptive coefficient (to be 1.35 for the surveys shown in [16]).

1. Based on the updating law (43) to filter the data samples satisfying condition (42)

$${}^{(r+1)}y_q\ \leftarrow\ {}^{(r)}y_q + \alpha\ \left|{}^{(r)}\left(y_q - \hat{y}_q\right)\right|\ \mathrm{sgn}\ \left({}^{(r)}\left(y_q - \hat{y}_q\right)\right)\ ,\ q = 1\dots\ \overline{Q}. \tag{43}$$

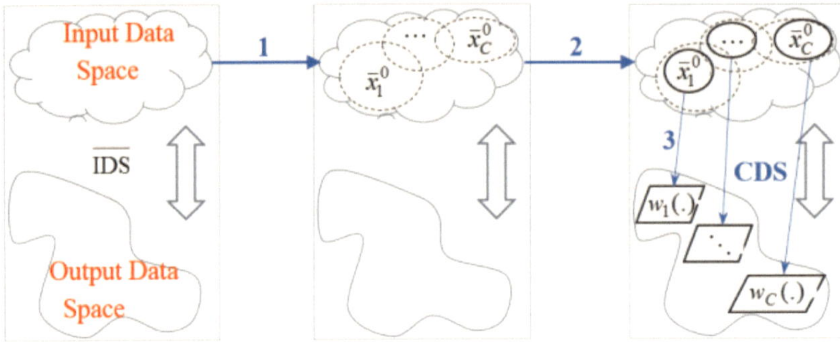

Figure 4.
A process of establishing the CDS driving from the $\overline{\text{IDS}}$ *consists.*

3.3 Algorithm for building ANFIS

Figure 4 illustrates the establishment of the CDS from the $\overline{\text{IDS}}$. It consists of (1) building fuzzy clusters with centroids $(\overline{x}_1^0, \ldots, \overline{x}_C^0)$ or the input data clusters (see Subsection 2.2), (2) estimating the hard distribution of samples in each input data cluster indicated by $(\overline{x}_1^0, \ldots, \overline{x}_C^0)$, and (3) building the hyperplanes or the output data clusters (see Subsection 2.3) in the output data space using the specified hard distribution status. Based on the created CDS, **Figure 3** shows the flowchart of the FIN-ANFIS consisting of three main phases: filtering IN, building the CDS driving from the filtered $\overline{\text{IDS}}$, and forming ANFIS.

3.4 Algorithm FIN-ANFIS

Initializing: The initial index of the loop process, $r = 1$; the number of clusters $C \ll P - 1$; $J_{KFCM}{}^{(r)} = \Omega$, where Ω is a real number $\Omega > [ts]$; and the initial cluster centroids corresponding to $r = 1$ chosen randomly:

$$\overline{x}_i^0(r) = \left(x_{i1}^0, \ldots, x_{in}^0\right), \quad 1 \le i \le C \tag{44}$$

Build the input data clusters:

1. Establish the input data clusters:

Based on the $\overline{x}_i^0(r)$ to be known, calculate μ_{ij} via Eq. (14) to update $\overline{x}_i^0(r)$ via Eq. (13).

2. Specify the stop condition of the clustering phase via ts in Eq. (15):

If $ts \le [ts]$: go to Step 3; ff $ts > [ts]$ and $r < [r]$, setup $r =: r + 1$ and return to Step 1; if $ts > [ts]$ and $r = [r]$ and $C < P - 1$, set $C =: C + 1$, $r =: 1$, and return to Step 1; and if $ts > [ts]$ and $r = [r]$ and $C = P - 1$, stop (not converge).

Build ANFIS:

3. Build and estimate ANFIS:

Establish ANFIS as presented in Subsection 2.4.

Calculate $\text{RMSE} = \left(P^{-1}\sum_{i=1}^{P}\left(\hat{y}_i(\overline{x}_i) - y_i\right)^2\right)^{0.5}$ in which $\hat{y}_i(\overline{x}_i)$ is the ANFIS-based output, while y_i is the data output. If $\text{RMSE} \le [E]$, stop (the ANFIS is the desired one); if $\text{RMSE} > [E]$ and $C < P - 1$, go to Step 4; and if $\text{RMSE} > [E]$ and $C = P - 1$, stop (not converge).

4. Set up a new cluster centroid:

Based on Eq. (41), seek the worst data point $WP \equiv \left(\overline{x}_i^{(WP)}, y_i^{(WP)}\right)$; set $C =: C + 1, r =: 1$, and set up a new cluster centroid $\overline{x}_C^0$ in the neighborhood of the WP; and go to Step 5.

5. Filter IN:

Call the algorithm AOINF and return to Step 1.

4. ANFIS for managing online bearing fault

An application of ANFIS to estimating online bearing fault upon the ability to extract meaningful information from big data of intelligent structures is shown in this section. Estimating online bearing status to hold the initiative in exploiting the systems is meaningful because bearing is an important machine detailed in almost mechanical structures.

In [17], an Online Bearing Damage Identifying Method (ASSBDIM) based on ANFIS, singular spectrum analysis (SSA), and sparse filtering (SF) was shown. The method consists of two phases: offline and online. In offline, the ANFIS identifies the dynamic response of the mechanical system in the individual bearing statuses. The trained ANFIS is then used to estimate its real status in the online phase. These aspects are detailed in the following paragraphs.

4.1 Some related theories

4.1.1 Singular spectrum analysis

By using SSA, from a given time series, a set of independent additive time series can be generated [41–43]. This work is clarified via the algorithm for SSA presented in [42] as follows.

1. Embedding:

Let's consider a given time series of N_O data points $(z_0, z_1, \ldots, z_{N_0-1})$. From selected window length L_O, $1 < L_O < N_O$, sliding vectors $\mathbf{X}^j = (z_{j-1}, z_j, \ldots, z_{j+L_0-2})^T$, $j = 1, \ldots, K = N_O - L_O + 1$, and matrix $\mathbf{X}$ as in Eq. (45) are built:

$$\mathbf{X} = \begin{pmatrix} z_0 & z_1 & \cdots & \cdots & z_{N_0-L_0} \\ z_1 & z_2 & \cdots & \cdots & z_{N_0-L_0+1} \\ \vdots & \vdots & \ddots & \therefore & \vdots \\ z_{L_0-2} & z_{L_0-1} & \therefore & \ddots & z_{N_0-2} \\ z_{L_0-1} & z_{L_0} & \cdots & \cdots & z_{N_0-1} \end{pmatrix}. \tag{45}$$

2. Building the trajectory matrix:

From Eq. (45), one builds matrix $\mathbf{S} = \mathbf{X}\mathbf{X}^T \in \Re^{L_0 \times L_0}$. Vectors $\mathbf{V}_i$ are then constructed, $\mathbf{V}_i = \mathbf{X}^T\mathbf{U}_i/\sqrt{\lambda_i}, \quad i = 1 \ldots d$, in which $\lambda_1, \ldots, \lambda_d$ are the non-zero eigenvalues of $\mathbf{S}$ arranged in the descending order and $\mathbf{U}_1, \ldots, \mathbf{U}_d$ are the corresponding eigenvectors. A decomposition of the trajectory matrix into a sum of matrices $\mathbf{X} = \sum_{i=1}^{d}\mathbf{E}_i$ is then established, where $\mathbf{E}_i = \sqrt{\lambda_i}\mathbf{U}_i\mathbf{V}_i^T$.

3. Reconstruction:

Each elementary matrix is transformed into a principal component of length N by applying a linear transformation known as diagonal averaging or Hankelization. Let $\mathbf{Z} \in \Re^{L_0 \times K}$ be a matrix of elements $z_{i,j}$.

By calculating $L^* = \min(L_0,\ K)$, $K^* = \max(L_0,\ K)$, then $\mathbf{Z}$ can be transformed into the reconstructed time series $g_0,\ g_1, \dots, g_{N-1}$ as in Eq. (46):

$$g_k = \begin{cases} \frac{1}{k+1}\sum_{m=1}^{k+1} z_{m,k-m+2}, & 0 \le k < L^* - 1 \\ \frac{1}{L^*}\sum_{m=1}^{L*} z_{m,k-m+2}, & L^* - 1 \le k < K^* \\ \frac{1}{N_0 - k}\sum_{m=k-K*+2}^{N-K*+1} z_{m,k-m+2}, & K^* \le k < N_0 \end{cases}. \tag{46}$$

4.1.2 Sparse filtering

In this work SF is used to extract features from a given time series-typed measured database. Relying an objective function defined via the features, the method tries to specify the good features such that the objective function is minimized [11, 44–45]. To deploy SF effectively, a process with the two following phases is operated. Preprocessing data based on the whitening method [46] is carried out in the first phase. A H-by-L matrix signed $\mathbf{F}$ of real numbers depicting the relation between each of the H training data samples and the L selected features is established in the second phase. SF presented in [11, 45] is detailed as follows.

In the first phase, a training set of the H data samples $\mathbf{x}_i \in \mathfrak{R}^{1\times N}$, $i = 1 \dots H$, in the form of a matrix signed $\breve{\mathbf{S}} \in \mathfrak{R}^{H\times N}$ is established from the given time series-typed measuring dataset. By adopting the whitening method [46], it then tries to make the data samples less correlated with each other and speed up the convergence of the sparse filtering process which employs the eigenvalue decomposition of the covariance matrix $\text{cov}\left(\breve{\mathbf{S}}\right) = \frac{\overline{\mathbf{Z}}}{\breve{\mathbf{D}}\overline{\mathbf{Z}}^T}$. In the expression, $\breve{\mathbf{D}}$ is the diagonal matrix of its eigenvalues, and $\overline{\mathbf{Z}}$ is the orthogonal matrix of eigenvectors of $\text{cov}\left(\breve{\mathbf{S}}\right)$. Finally, the whitened training set signed $\mathbf{S}_{white}$ is formed as in Eq. (47):

$$\mathbf{S}_{white} = \frac{\overline{\mathbf{Z}}}{\breve{\mathbf{D}}-1/2}\overline{\mathbf{Z}}^T\breve{\mathbf{S}}. \tag{47}$$

Subsequently, in the second phase, SF maps the data sample $\mathbf{x}_i \in \mathfrak{R}^{1\times N}$ of $\mathbf{S}_{white}$ onto L features f_i, $i = 1 \dots L$, relied on a weight matrix signed $\overline{\mathbf{W}} \in \mathfrak{R}^{N\times L}$. A linear relation between data samples in $\mathbf{S}_{white}$ and the L features is expressed via $\overline{\mathbf{W}}$ as in Eq. (48), in which $\mathbf{F} \in \mathfrak{R}^{H\times L}$ is called the feature distribution matrix:

$$\mathbf{F} = \mathbf{S}_{white}\overline{\mathbf{W}}. \tag{48}$$

Optimizing the feature distribution in $\mathbf{F}$ is then performed as detailed in [45]. The features in each column of $\mathbf{F}$ is normalized by dividing them by their l_2-norm, $\tilde{\mathbf{f}}l = \mathbf{f}^l / \left\|\mathbf{f}^l\right\|_2$, $l = 1 \dots L$. For each row of the obtained matrix, these features per example are normalize by computing $\hat{\mathbf{f}}_i = \tilde{\mathbf{f}}_i / \left\|\tilde{\mathbf{f}}_i\right\|_2$, $i = 1 \dots H$, by which they lie on the unit l_2-ball. The features normalized after the two above steps are optimized for sparseness using the l_1-penalty to get a matrix signed $\hat{\mathbf{F}} \in \mathfrak{R}^{H\times L}$. A loop process is then maintained via Eq. (48), in which $\hat{\mathbf{F}}$ takes the role of $\mathbf{F}$, until the optimal weights of $\overline{\mathbf{W}}$ are to be established that make the objective function $J_{SF}\left(\overline{\mathbf{W}}\right)$ of Eq. (49) be minimized, to which, finally, $\hat{\mathbf{F}}$ is resigned $\mathbf{F}$:

$$J_{SF}\left(\overline{\mathbf{W}}\right) = \sum_{i=1}^{H}\sum_{j=1}^{L}\hat{\mathbf{F}}(i,j). \tag{49}$$

4.2 The ASSBDIM

The ASSBDIM focuses on online bearing fault estimation. The aim is detailed in this subsection consisting of the way of setting up the databases and the algorithm ASSBDIM for online bearing fault estimation upon the built databases.

4.2.1 Building the databases for the ASSBDIM

A measuring dataset deriving from the mechanical system vibration is established for each surveyed bearing fault type. Regarding Q fault types, one obtains Q original datasets as in Eq. (50):

$$\left[\mathbf{D}_1,\ \mathbf{D}_2, \ldots, \mathbf{D}_Q\right]^T, \tag{50}$$

where $\mathbf{D}_i$ is corresponding to the ith bearing fault type $(1 \le i \le Q)$.

By using SSA for $\mathbf{D}_i$, m time series as in Eq. (51) are set up:

$$[\mathbf{D}_{i1},\ \mathbf{D}_{i2}, \ldots, \mathbf{D}_{im}], i = 1 \ldots Q \tag{51}$$

where m is parameter selected by the designer. This work is carried out by the three steps as presented in Subsection 4.1.1, in which $\mathbf{D}_i$ is used in the first step as the given time series of N_0 data points $(z_0,\ z_1, \ldots, z_{N_0-1})$ for building the trajectory matrix $\mathbf{X}$ in Eq. (45). Because the mechanical vibration signal is prone to the low frequency range [42], among the m time series, the $(m\text{-}k)$ ones owning the highest frequencies are considered as noise. The k remainder time series as in Eq. (52) is hence kept to build the databases:

$$[\mathbf{D}_{i1},\ \mathbf{D}_{i2} \ldots \mathbf{D}_{ik}], i = 1 \ldots Q \tag{52}$$

Specifying the optimal value of both k and m will be mentioned in Subsection 4.2.2.

For each time series in Eq. (52), for example, $\mathbf{D}_{ij}, j = 1 \ldots k$, based on SF one obtains the feature distribution matrix as in Eq. (48) which is resigned $\mathbf{F}_{ij}(\omega) \in \Re^{H \times L}$. By using this result for all the time series in Eq. (52), a new data matrix $\overline{\mathbf{D}}_i$ as in Eq. (53) is formed which is the input data space of the ith data subset corresponding to the ith bearing fault type:

$$\overline{\mathbf{D}}_i = [\mathbf{F}_{i1}(\omega)\, \mathbf{F}_{i2}(\omega) \ldots \mathbf{F}_{ik}(\omega)] \in \Re^{H \times (kL)}. \tag{53}$$

By employing this way for Q, the surveyed bearing fault types, an input data space in the form of matrix (54), are established, which relates to building two offline databases signed Off_DaB and Off_testDaB as well as one online database signed On_DaB used for the algorithm ASSBDIM as follows:

$$\overline{\mathbf{D}} = \begin{pmatrix} \omega_{11} & \omega_{12} & \cdots & \cdots & \omega_{1(kL)} \\ \omega_{21} & \omega_{22} & \cdots & \cdots & \omega_{2(kL)} \\ \vdots & \vdots & \ddots & ⋰ & \vdots \\ \omega_{(QH-1)1} & \omega_{(QH-1)2} & ⋰ & \ddots & \omega_{(QH-1)(kL)} \\ \omega_{(QH)1} & \omega_{(QH)2} & \cdots & \cdots & \omega_{(QH)(kL)} \end{pmatrix} \in \Re^{(QH) \times (kL)} \tag{54}$$

Namely, matrix $\overline{\mathbf{D}}$ relates to the input data space (IDS), to which the databases for identifying the bearing status are built as follows. Firstly, by encoding the ith fault type by a real number y_i, the output data space (ODS) of the ith subset can be depicted by vector $\overline{\mathbf{y}}_i$ of H elements y_i as in Eq. (55):

$$\overline{\mathbf{y}}_i = \left[y_i, ..., y_i\right]^T \in \mathfrak{R}^{H\times 1}, i = 1...Q \tag{55}$$

Then, by combining Eq. (55) with Eq. (54), the input-output relation in the three datasets Off_DaB, Off_testDaB, and On_DaB can be described as in Eq. (56):

$$\text{database} \equiv [\text{IDS} - \text{ODS}] \equiv \left[\overline{\mathbf{D}} - \overline{\mathbf{y}}\right] \tag{56}$$

In the above, the input space $\overline{\mathbf{D}}$ comes from Eq. (54), while the output space $\overline{\mathbf{y}}$ as in Eq. (57) is constituted of $\overline{\mathbf{y}}_i \in \mathfrak{R}^{H\times 1}$ in Eq. (55):

$$\overline{\mathbf{y}} = \left[y_1, ...y_{1H}, ..., y_Q, ...y_{QH}\right]^T \in \mathfrak{R}^{QH\times 1} \tag{57}$$

4.2.2 The algorithm ASSBDIM for estimating health of bearings

In the offline phase, by initializing the parameters in vector **ps** in Eq. (58), together with applying SSA and SF to the measuring data stream, the Off_DaB and Off_testDaB are built as in Eq. (56):

$$\mathbf{ps} = [L_0, N_0, m, k, H, L] \tag{58}$$

where L_0, N_0 come from Eq. (45); m and k relate to Eqs. (51) and (52), respectively; H and L derive from Eq. (53).

An ANFIS built by the algorithm FIN-ANFIS (see Subsection 3.3) is utilized to identify dynamic response of the mechanical system corresponding to the bearing damage statuses reflected by the Off_DaB. Optimizing the parameters in **ps** in Eq. (58) is then performed using the percentage of correctly estimated samples (*Ac*) as in Eq. (59) and the mean accuracy (*MeA*) as in Eq. (60) and the algorithm DE [47]:

$$Ac = 100 \times cr\_samples_n / to\_samples_n (\%), \tag{59}$$

$$MeA = 100 \times \sum_{n=1}^{Q} cr\_samples_n / \sum_{n=1}^{Q} to\_samples_n (\%), \tag{60}$$

where corresponding to the nth damage type, $n = 1...Q$, $cr\_samples_n$ is the number of checking samples expressing correctly the real status of the bearing, while $to\_samples_n$ is the total of checking samples used in the survey; Q is the number of surveyed bearing fault types as mentioned in Eq. (50).

Following the *MeA*, an objective function is defined as follows:

$$J = MeA_{\text{ASSBDIM}}(L_0, N_0, m, k, H, L) \rightarrow \max. \tag{61}$$

The Off_testDaB, function J, and DE [47] are then employed to optimize the parameters in vector **ps**, to get $[L_0, N_0, m, k, H, L]_{opt}$.

Namely, by using the input of the Off_testDaB for the ANFIS which has been trained by the Off_DaB, one obtains the outputs $\hat{y}_i$, $i = 1...H$. These outputs are then compared with the corresponding encoded outputs to estimate the bearing real status, which is the one encoded by "q" satisfying Eq. (62):

$$\sum_{i=1}^{H}\left|\hat{y}_i - y_q\right| = \min_{h=1\ldots Q} \sum_{i=1}^{H}\left|\hat{y}_i - y_h\right|. \tag{62}$$

The completion of the offline phase as above can be seen as the beginning of the only phase. During the next operating process, first, by the way similar to the one for building the offline database Off_DaB, the online dataset On_DaB in the form $\overline{\mathbf{D}}_{\mathrm{ON}} \equiv \overline{\mathbf{D}}_i \in \Re^{H\times(kL)}$ as in Eq. (53) is built. By using the On_DaB for the ANFIS trained in the offline, the bearing real status at this time is then specified based on Eq. (62).

The ASSBDIM is hence can be summarized as follows.

The offline process:
Initialize vector **ps** in Eq. (58):

1. Build the Off-DaB and Off-testDaB in the form of Eq. (56).
2. Train an ANFIS to identify the Off-DaB using the algorithm FIN-ANFIS.
3. Accomplish the system.

The Off-testDaB is used as database of the trained ANFIS, using the condition (62) to calculate MeA in Eq. (60). If $MeA \le [MeA]$, then go to Step 4; otherwise, adjust the value of the elements in vector **ps** in Eq. (58) using the algorithm DE [47], and then return to Step 1.

The online process:

4. Establish online database On-DaB $\overline{\mathbf{D}}_{\mathrm{ON}} \equiv \overline{\mathbf{D}}_i \in \Re^{H\times(kL)}$ as in Eq. (53).
5. Estimate online bearing fault status based on the On-DaB, trained ANFIS, and condition (62); check the stop condition: if it is not satisfied, then return to Step 4; otherwise, stop.

4.3 Some survey results

4.3.1 Experimental apparatus and estimating way

The experimental apparatus for measuring vibration signal is shown in **Figure 5**. The apparatus consists of the motor (1), acceleration sensors (2) and (4), surveyed bearings (3) and (5), module for processing and transforming series vibration signal incorporating software-selectable AC/DC coupling (Model: NI-9234) (6), and computer (7).

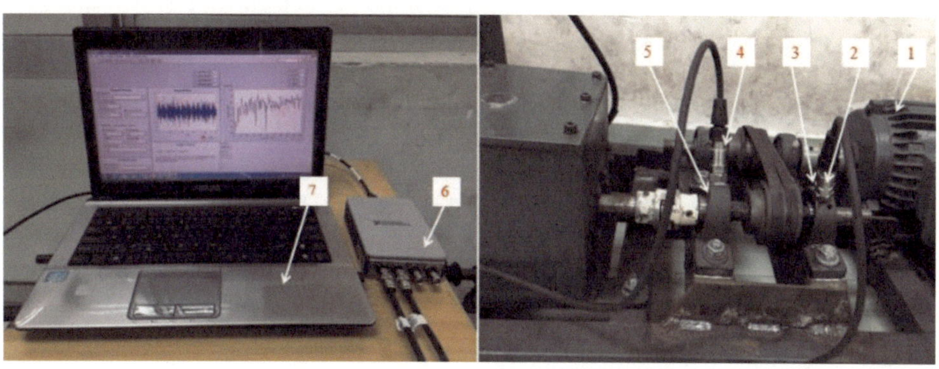

Figure 5.
Experimental apparatus for measuring vibration signal.

Surveyed cases					
Case 1		Case 2		Case 3	
Status	EV (y_i)	Status	EV (y_i)	Status	EV (y_i)
L1Und	0	L2Und	0	L1Und	0
L1D1In	1	L2D1In	1	L1D1Ba	1
L1D2In	2	L2D2In	2	L1D1In	2
L1D3In	3	L2D3In	3	L1D1Ou	3
L1D1Ou	4	L2D1Ba	4	L1D2Ba	4
L1D2Ou	5	L2D2Ba	5	L1D2In	5
L1D3Ou	6	L2D3Ba	6	L1D2Ou	6
-	-	-	-	L1D3Ba	7
-	-	-	-	L1D3In	8
-	-	-	-	L1D3Ou	9

Table 1.
Surveyed cases and the corresponding encoding values (EV).

Faults	Width (mm)	Depth (mm)
D1Ou	0.20	0.3
D2Ou	0.30	0.3
D3Ou	0.46	0.3
D1In	0.20	0.3
D2In	0.30	0.3
D3In	0.40	0.3
D1Ba	0.15	0.2
D2Ba	0.20	0.2
D3Ba	0.25	0.2

Table 2.
The size of bearing single fault types used for surveys.

In **Table 1**, "encoding value" is abbreviated to "EV." The three cases listed in **Table 1** related to nine of the widespread single-bearing faults as in **Table 2** are surveyed. In the above, $Q = 7$ (see Eq. 50) for the Cases 1–2, while $Q = 10$ for Case 3; the damaged location is the inner or outer or balls (signed In, or Ou, or Ba, respectively); damaged degrees are from 1 to 3 (signed D1 or D2 or D3); the load impacting on the system at the survey time consists of Load 1 or 2 or 3 (signed L1 or L2 or L3). For example, LmUnd shows the load degree to be m and the bearing to be undamaged, or LmDnBa expresses the load degree to be m (1,...,3), the damage level to be n (1,...,3), and the damage location to be the ball.

The ASSBDIM with $H = 303$, $m = 30$, $k = 7$ along with four other methods [48–51] is employed to be surveyed. The first one [48] ($N_{in} = N_{out} = 100$; number of segments to be 20×10^3 and $\lambda = 1E - 5$) is the intelligent fault diagnosis method using unsupervised feature learning toward mechanical big data. The second one [49] employs the energy levels of the various frequency bands as features. In the third one [50], a bearing fault diagnosis upon permutation entropy, empirical mode decomposition, and support vector machines is shown. In the last one [51], a method of identifying bearing fault based on SSA is presented.

For the surveys, along with Ac and MeA, the root-mean-square error as in Eq. (63) is also employed, where y_i and $\hat{y}_i$, respectively, are encoding and predicting outputs:

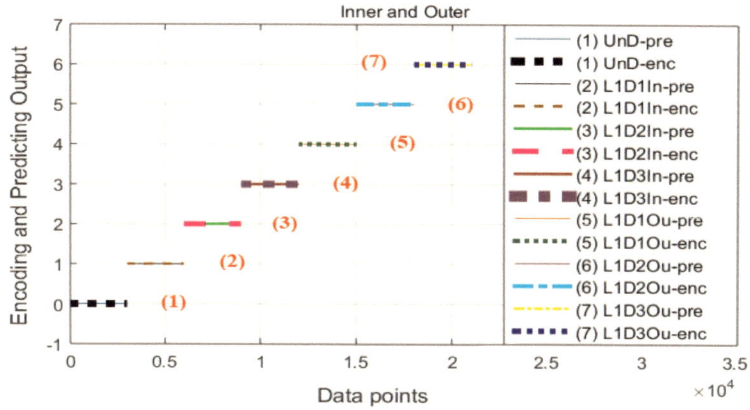

Figure 6.
The predicting (−pre) output $\hat{y}_i$ of the ASSBDIM in Case 1 and the corresponding encoded (−enc) output y_i.

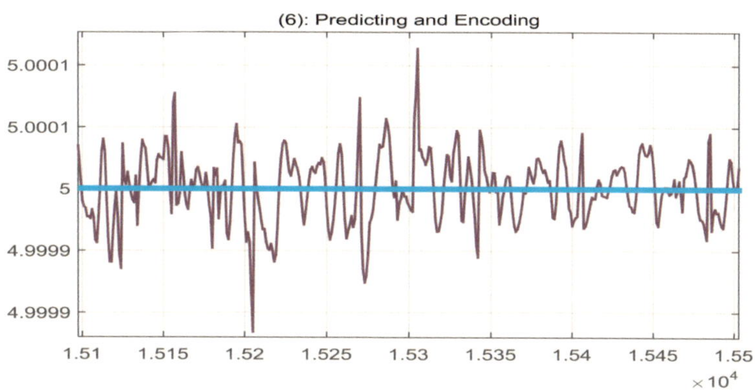

Figure 7.
The $\hat{y}_i$ and y_i depicted by lines (6) in ***Figure 6*** *to be zoomed in.*

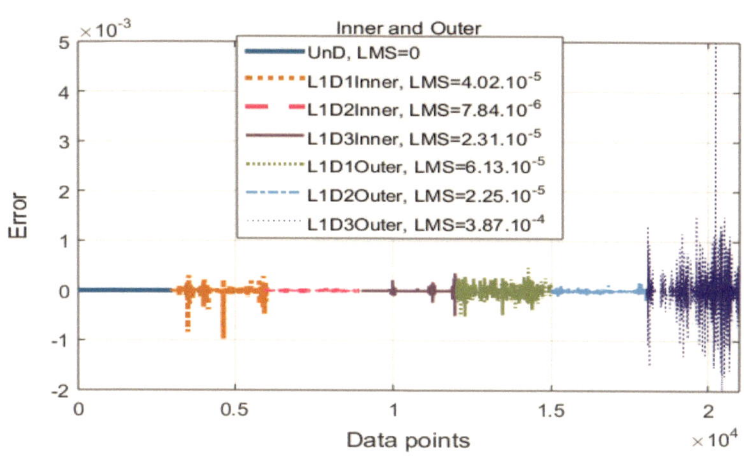

Figure 8.
The error reflecting the difference between y_i and $\hat{y}_i$ in ***Figure 6***.

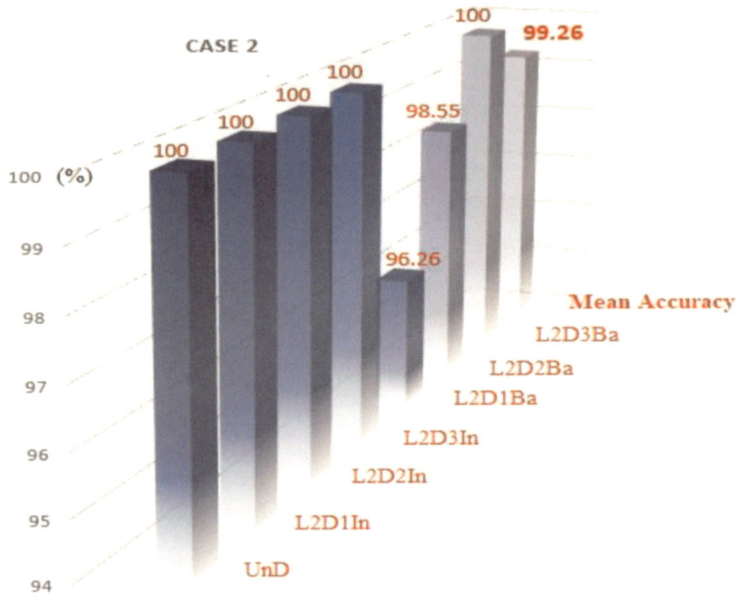

Figure 9.
Ac and MeA (mean accuracy) of the ASSBDIM in Case 2.

Surveyed cases	*Ac* (%)				
	[48]	[49]	[50]	[51]	[17]
L_2UnD	99.67	93.73	98.68	99.67	100
L_2D_1In	98.35	95.05	92.74	95.38	99.67
L_2D_2In	99.67	98.68	99.34	97.36	99.01
L_2D_3In	99.01	93.07	95.38	92.74	100
L_2D_1Ba	98.68	91.09	96.37	96.70	97.36
L_2D_2Ba	99.67	92.08	94.72	99.01	98.68
L_2D_3Ba	97.36	98.68	100	98.68	100
MeA (%)	98.92	94.93	96.75	97.08	**99.26**

Table 3.
The accuracy of the methods in Case 2.

$$\mathrm{LMS} = \sqrt{\sum_{i=1}^{H}(y_i - \hat{y}_i)^2/H}. \tag{63}$$

4.3.2 Some survey results

The measured databases from Cases 1 to 3 with $Q = 7$ as in **Table 1** along which the methods consist of the ASSBDIM [17] and the ones from [48–51] were adopted to identify the status of the bearing. The obtained results were shown in **Figures 6–9** and **Tables 3** and **4**.

4.3.3 Discussion

Following the above results, it can observe that among the surveyed methods, the ASSBDIM which is based on ANFIS gained the best accuracy. This aspect can be

Surveyed cases	*Ac* (%)				
	[48]	[49]	[50]	[51]	[17]
L_1UnD	95.05	87.46	85.81	100	100
L_1D_1In	94.72	90.10	89.77	83.17	99.67
L_1D_2In	92.08	92.41	88.12	85.48	99.34
L_1D_3In	93.40	92.74	92.41	94.72	99.34
L_1D_1Ou	95.33	89.77	84.16	85.15	82.51
L_1D_2Ou	92.41	92.41	84.82	84.16	94.39
L_1D_3Ou	95.05	88.78	88.78	99.34	89.11
L_1D_1Ba	86.47	89.44	90.43	83.17	94.72
L_1D_2Ba	87.79	90.10	96.04	97.36	92.74
L_1D_3Ba	88.12	86.14	100	88.45	84.82
MeA (%)	92.04	89.94	90.03	90.10	**93.66**

Table 4.
The accuracy of the methods in Case 3.

recognized via the quite equivalent values between the encoding and predicting outputs from the tested data samples. The small difference depicted by the zooming in in **Figure 7** and the root-mean-square error in **Figure 8** as well as the high/higher values of Ac and MeA deriving from the ASSBDIM in **Tables 3** and **4** and **Figure 9** reflect clearly the ANFIS's identification ability.

It should be noted that the methodology shown via the algorithm ASSBDIM can be also used to discover the method of managing damage of mechanical structures as well.

5 Conclusion

The hybrid structure ANFIS, where ANN and FL can interact to not only overcome partly the limitations of each model but also uphold their strong points, has been seen as a useful mathematical tool for many fields. Inspired by the ANFIS's capability, in order to provide the readers with the theoretical basis and application direction of the model, this chapter presents the formulation of ANFIS and one of its typical applications.

Firstly, the structure of ANFIS as a data-driven model deriving from fuzzy logic and artificial neural networks is depicted. The setting up the input data clusters, output clusters and ANFIS as a joint structure is all detailed. Deriving from this relation, the method of building ANFIS from noisy measuring datasets is presented. The online and recurrent mechanism for filtering noise and building ANFIS synchronously is clarified via the algorithms for filtering noise and establishing ANFIS. Finally, the application of ANFIS coming from the online managing bearing fault is presented. The compared results reflect that among the surveyed methods, the ASSBDIM which exploited the identification ability of ANFIS gains the best accuracy. Besides, the methodology shown via this application can be also used as appropriate solution for developing new methods of managing damage of mechanical structures.

In addition to the above identification field, it should be noted that (1) ANFIS has also attracted the attention of many researchers in the other fields related to prediction, control, and so on, as mentioned in Section 1 and (2) ANFIS can collaborate effectively with some other mathematical tools to enhance the effectiveness of technology applications.

Author details

Sy Dzung Nguyen

1 Division of Computational Mechatronics, Institute for Computational Science, Ton Duc Thang University, Ho Chi Minh City, Vietnam

2 Faculty of Electrical and Electronics Engineering, Ton Duc Thang University, Ho Chi Minh City, Vietnam

*Address all correspondence to: nguyensydung@tdtu.edu.vn

References

[1] Kosko B. Fuzzy systems as universal approximators. IEEE Transactions on Computers. 1994;**43**(11):1329-1333

[2] Nguyen SD, Choi SB. A new neuro-fuzzy training algorithm for identifying dynamic characteristics of smart dampers. Smart Materials and Structures. 2012;**21**(8):1-14

[3] Nguyen SD, Choi SB. A novel minimum-maximum data-clustering algorithm for vibration control of a semi-active vehicle suspension system. Journal of Automobile Engineering, Part D. 2013;**227**(9):1242-1254

[4] Nguyen SD, Choi SB, Nguyen QH. An optimal design of interval type-2 fuzzy logic system with various experiments including magnetorheological fluid damper. Proceedings of the Institution of Mechanical Engineers, Part C: Journal of Mechanical Engineering Science. 2014: 1-17. DOI: 10.1177/0954406214526585

[5] Nguyen SD, Nguyen QH, Choi SB. Hybrid clustering based fuzzy structure for vibration control – Part 1: A novel algorithm for building neuro-fuzzy system. In: Mechanical Systems and Signal Processing. Vol. 50-51. 2014. pp. 510-525

[6] Nguyen SD, Choi SB. Design of a new adaptive neuro-fuzzy inference system based on a solution for clustering in a data potential field. Fuzzy Sets and Systems. 2015;**279**:64-86

[7] Nguyen SD, Nguyen QH. Design of active suspension controller for train cars based on sliding mode control, uncertainty observer and neuro-fuzzy system. Journal of Vibration and Control. 2015:1-20. DOI: 10.1177/1077546315592767

[8] Jang JSR. ANFIS: Adaptive-network-based fuzzy inference systems. IEEE Transactions on Systems, Man, and Cybernetics. 1993;**23**:665-685

[9] Chen C, Bin Z, George V, Marcos O. Machine condition prediction based on adaptive neuro-fuzzy and high-order particle filtering. IEEE Transactions on Industrial Electronics. 2011;**58**(9): 4353-4364

[10] Panella M, Gallo AS. An input–output clustering approach to the synthesis of ANFIS networks. IEEE Transactions on Fuzzy Systems. 2005; **13**(1):69-81

[11] Lei Y, Jia F, Lin J, Xing S, Ding SX. An intelligent fault diagnosis method using unsupervised feature learning towards mechanical big data. IEEE Transactions on Industrial Electronics. 2016;**63**(5)

[12] Nguyen SD, Nguyen QH, Seo TI. ANFIS deriving from jointed input-output data space and applying in smart-damper identification. Applied Soft Computing. 2017;**53**:45-60

[13] Theocharis JB. A high-order recurrent neuro-fuzzy system with internal dynamics: Application to the adaptive noise cancellation. Fuzzy Sets and Systems. 2006;**157**:471-500

[14] Besdok E, Civicioglu P, Alci M. Using an adaptive neuro-fuzzy inference system based interpolant for impulsive noise suppression from highly distorted images. Fuzzy Sets and Systems. 2005;**150**:525-543

[15] Kumari R, Gambhir D, Kumar V, Intensity difference based neuro-fuzzy system for impulse noisy image restoration: ID-NFS. In: Proceedings of International Conference on Signal Processing and Integrated Networks (SPIN), 978-1-4799-2866-8/14/$31.00 ©2014 IEEE, 2014

[16] Nguyen SD, Choi S-B, Seo T-I. Recurrent mechanism and impulse noise filter for establishing ANFIS. IEEE Transactions on Fuzzy Systems. 2018; **26**(2):985-997

[17] Seo T-I, Sy DN. Algorithm for estimating online bearing fault upon the ability to extract meaningful information from big data of intelligent structures. IEEE Transactions on Industrial Electronics. 2018. DOI: 10.1109/TIE.2018.2847704

[18] Nguyen SD, Choi SB, Nguyen QH. A new fuzzy-disturbance observer-enhanced sliding controller for vibration control of a train-car suspension with magneto-rheological dampers. Mechanical Systems and Signal Processing. 2018;**105**:447-466

[19] Nguyen SD, Jung D, Choi SB. A robust vibration control of a magnetorheological damper based railway suspension using a novel adaptive type-2 fuzzy sliding mode controller. Shock and Vibration. 2017: 7306109. DOI: 10.1155/2017/7306109

[20] Nguyen SD, Vo HD, Seo TI. Nonlinear adaptive control based on fuzzy sliding mode technique and fuzzy-based compensator. ISA Transactions. 2017;**70**:309-321

[21] Nguyen SD, Ho HV, Nguyen TT, Truong NT, Seo TI. Novel fuzzy sliding controller for MRD suspensions subjected to uncertainty and disturbance. Engineering Applications of Artificial Intelligence. 2017;**61**:65-76

[22] Nguyen SD, Nguyen QH, Seo TI. ANFIS deriving from jointed input-output data space and applying in smart-damper identification. Applied Soft Computing. 2017;**53**:45-60

[23] Nguyen SD, Kim WH, Park JH, Choi SB. A new fuzzy sliding mode controller for vibration control systems using integrated structure smart dampers. Smart Materials and Structures. 2017; **26**(2017):045038

[24] Nguyen SD, Choi SB, Seo TI. Adaptive fuzzy sliding control enhanced by compensation for explicitly unidentified aspects. International Journal of Control, Automation and Systems. 2017. DOI: 10.1007/s12555-016-0569-6

[25] Nguyen SD, Seo TI. Establishing ANFIS and the use for predicting sliding control of active railway suspension systems subjected to uncertainties and disturbances. International Journal of Machine Learning and Cybernetics. 2016. DOI: 10.1007/s13042-016-0614-z

[26] Nguyen SD, Nguyen QH. Design of active suspension controller for train cars based on sliding mode control, uncertainty observer and neuro-fuzzy system. Journal of Vibration and Control. 2015;**23**(8):1334-1353

[27] Turkmen I. Efficient impulse noise detection method with ANFIS for accurate image restoration. International Journal of Electronics and Communications. 2011;**65**:132-139

[28] Hemalatha C, Periasamy A, Muruganand S. A hybrid approach for efficient removal of impulse, Gaussian and mixed noise from highly corrupted images using adaptive neuro fuzzy inference system (ANFIS). International Journal of Computer Applications. 2012; **45**(16):15-21

[29] Saradhadevi V, Sundaram DV. An enhanced two-stage impulse noise removal technique based on fast ANFIS and fuzzy decision. International Journal of Computer Science Issues. 2011;**8**(1):79-88

[30] Shen H, Yang J, Wangm S, Liu X. Attribute weighted mercer kernel based fuzzy clustering algorithm for general non-spherical datasets. Soft Computing. 2006;**10**:1061-1073

[31] Marcelo RP, Ferreira AT, Francisco C. Kernel fuzzy C-means with automatic variable weighting. Fuzzy Sets and Systems. 2014;**237**:1-46

[32] Filippone M, Camastra F, Masulli F, Rovetta S. A survey of kernel and spectral methods for clustering. Pattern Recognition. 2008;**41**:176-190

[33] Camastra F, Verri A. A novel kernel method for clustering. IEEE Transactions on Neural Networks. 2005; **27**(5):801-804

[34] Graves D, Pedrycz W. Kernel-based fuzzy clustering and fuzzy clustering: A comparative experimental study. Fuzzy Sets and Systems. 2010;**161**:522-543

[35] Winkler R, Klawonn F, Kruse R. Problems of fuzzy c-means clustering and similar algorithms with high dimensional data sets. International Journal of Fuzzy Systems. 2011;**1**(1):1-16

[36] Xu R, Wunusch DII. Survey of clustering algorithms. IEEE Transactions on Neural Networks. 2005; **16**(3):645-678

[37] Girolami M. Mercer kernel-based clustering in feature space. IEEE Transactions on Neural Networks. 2002; **13**:780-784

[38] Thevaril J, Kwan HK. Speech enhancement using adaptive neuro-fuzzy filtering. Proceedings of International Symposium on Intelligent Signal Processing and Communication Systems. 2005

[39] Balaiah P, Ilavennila. Comparative evaluation of adaptive filter and neuro-fuzzy filter in artifacts removal from electroencephalogram signal. American Journal of Applied Sciences. 2012;**9**(10): 1583-1593

[40] Lakra S, Prasad TV, Ramakrishna G. Selective noise filtering of speech signals using an adaptive neuro-fuzzy inference system as a frequency pre-classifier. Journal of Theoretical and Applied Information Technology. 2015; **81**(3):496-501

[41] Golyandina, Nekrutkin V, Zhigljavsky A. Analysis of Time Series Structure—SSA and Related Techniques, Chapman & Hall/CRC. Boca Raton, Florida; 2001

[42] Salgado DR, Alonso FJ. Tool wear detection in turning operations using singular spectrum analysis. Journal of Materials Processing Technology. 2006; **171**:451-458

[43] Kilundu B, Dehombreux P, Chiementin X. Tool wear monitoring by machine learning techniques and singular spectrum analysis. Mechanical Systems and Signal Processing. 2011;**25**: 400-415

[44] Willmore B, Tolhurst DJ. Characterizing the sparseness of neural codes. Network: Computation in Neural Systems. 2001;**12**(3):255-270

[45] Ngiam J, Chen Z, Bhaskar SA, Koh PW, Ng AY. Sparse filtering. In: Proceedings of Neural Information Processing Systems. 2011. pp. 1125-1133

[46] Hyvärinen A, Oja E. Independent component analysis: Algorithms and applications. Neural Networks. 2000; **13**(4):411-430

[47] Gong W, Cai Z. Differential evolution with ranking-based mutation operators. IEEE Transactions on Cybernetics. 2013;**43**:1-16

[48] Lei Y, Jia F, Lin J, Xing S, Ding SX. An intelligent fault diagnosis method using unsupervised feature learning towards mechanical big data. IEEE Transactions on Industrial Electronics. 2016;**63**(5)

[49] Ao HL, Cheng J, Li K, Truong TK, A Roller Bearing Fault Diagnosis Method

Based on LCD Energy Entropy and ACROA-SVM, Shock and Vibration; 2014. Vol. 2014. Article ID: 825825. 12 p

[50] Zhang X, Liang Y, Zhou J, Zang Y. A novel bearing fault diagnosis model integrated permutation entropy, ensemble empirical mode decomposition and optimized SVM. Measurement. 2015;**69**:164-179

[51] Liu T, Chen J, Dong G. Singular spectrum analysis and continuous hidden Markov model for rolling element bearing fault diagnosis. Journal of Vibration and Control. 2015;**21**(8): 1506-1521

Section 4

Inference Methods

Chapter 4

Some Methods of Fuzzy Conditional Inference for Application to Fuzzy Control Systems

Poli Venkata Subba Reddy

Abstract

Zadeh proposed fuzzy logic with single membership function. Two Zadeh, Mamdani and TSK proposed fuzzy conditional inference. In many applications like fuzzy control systems, the consequent part may be derived from precedent part. Zadeh, Mamdani and TSK proposed different fuzzy conditional inferences for "if … then …" for approximate reasoning. The Zadeh and Mamdani fuzzy conditional inferences are know prior information for both precedent part and consequent part. The TSK fuzzy conditional inferences need not know prior information for consequent part but it is difficult to compute. In this chapter, fuzzy conditional inference is proposed for "if…then…" This fuzzy conditional inference need not know prior information of the consequent part. The fuzzy conditional inference is discussed using the single fuzzy membership function and twofold fuzzy membership functions. The fuzzy control system is given as an application.

Keywords: fuzzy logic, twofold fuzzy logic, fuzzy conditional inference, fuzzy control systems

1. Introduction

When information is incomplete, fuzzy logic is useful [10–26]. Many theories [1, 2] deal with incomplete information based on likelihood (probability), whereas fuzzy logic is based on belief. Zadeh defined fuzzy set with single membership function. Zadeh [3], Mamdani [4], TSK [2] and Reddy [5] are studied fuzzy conditional inferences. The fuzzy conditions are of the form "if <. Zadeh, Mamdani and TSK fuzzy conditional inference requires both precedent-part and consequent-part but 5fuzzy inferences don't require consequent part. Precedent-part > then <consequent-part >."

Zadeh [6] studied fuzzy logic with single membership function. The single membership function for the proposition "x is A" contains how much truth in the proposition. The fuzzy set with two membership functions will contain more information in terms of how much truth and false it has in the proposition. The fuzzy certainty factor is studied as difference on two membership functions "true" and "false" to eliminate conflict of evidence, and it becomes single membership function. The FCF is a fuzzy set with single fuzzy membership function of twofold fuzzy set.

The fuzzy control systems are considered in this chapter as application of single fuzzy membership function and twofold fuzzy set.

2. Fuzzy log with single membership function

Zadeh [6] has introduced a fuzzy set as a model to deal with imprecise, inconsistent and inexact information. The fuzzy set is a class of objects with a continuum of grades of membership.

The fuzzy set A of X is characterized as its membership function $A = \mu_A(x)$ and ranging values in the unit interval [0, 1]

$\mu_A(x): X \rightarrow [0, 1], x \in X$,

where X is the universe of discourse.

$A = \mu_A(x_1)/x1 + \mu_A(x_2)/x_2 + \ldots + \mu_A(x_n)/x_n$,

where "+" is the union.

For instance, the fuzzy proposition "x is High"

$High = 0.2/x_1 + 0.6/x_2 + 0.9/x_3 + 0.6/x_4 + 0.2/x_5$

$Not\ High = 0.8/x_1 + 0.4/x_2 + 0.1/x_3 + 0.4/x_4 + 0.8/x_5$

For instance, the fuzziness of "Temperature is high" is 0.8

The graphical representation of young and not young is shown in **Figure 1**.

The fuzzy logic is defined as a combination of fuzzy sets using logical operators. Some of the logical operations are given below.

For example, A, B and C are fuzzy sets. The operations on fuzzy sets are given as:

Negation

If x is not A

$A' = 1 - \mu_A(x)/x$

Conjunction

x is A and y is B→ (x, y) is A ΛB

$A \Lambda B = \min(\mu_A(x), \mu_B(y)\}(x,y)$

If x = y

x is A and y is B→ (x, y) is A ΛB

$A \Lambda B = \min(\mu_A(x), \mu_B(y)\}/x$

For example

$A = 0.2/x_1 + 0.6/x_2 + 0.9/x_3 + 0.6/x_4 + 0.2/x_5$

$B = 0.4/x_1 + 0.6/x_2 + 0.9/x_3 + 0.6/x_4 + 0.1/x_5$

$A \Lambda B = 0.2/x_1 + 0.6/x_2 + 0.9/x_3 + 0.6/x_4 + 0.1/x_5$

The graphical representation is shown in **Figures 1** and **2**.

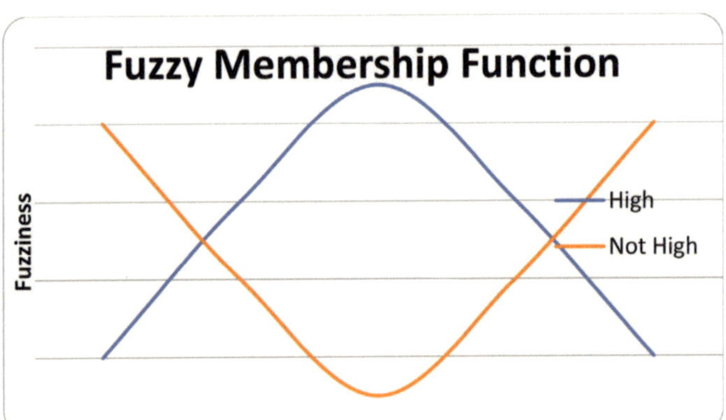

Figure 1.
Fuzzy membership function.

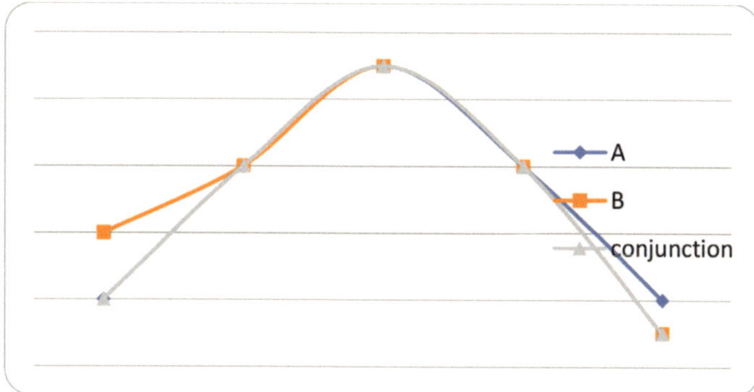

Figure 2.
Conjunction.

Disjunction

x is A and y is B→ (x, y) is A V B

A V B = max(μ_A(x), μ_B(y)}(x,y)

If x = y

x is A and y is B→ (x, y) is A V B

AVB = max(μ_A(x), μ_B(y)}/x

For instance,

$A = 0.2/x_1 + 0.6/x_2 + 0.9/x_3 + 0.6/x_4 + 0.2/x_5$

$B = 0.4/x_1 + 0.6/x_2 + 0.9/x_3 + 0.6/x_4 + 0.1/x_5$

$AVB = 0.4/x_1 + 0.6/x_2 + 0.9/x_3 + 0.6/x_4 + 0.2/x_5$

The graphical representation is shown in **Figure 3**.

Concentration

$\mu_{\text{very } A}(x) = \mu_A(x)^2$

Diffusion

$\mu_{\text{more or less } A}(x) = \mu_A(x)^{0.5}$

The graphical representation of concentration and diffusion is shown in **Figure 4**.

Implication

Zadeh [6], Mamdani [7] and Reddy [5] fuzzy conditional inferences are considered for fuzzy control systems.

If x_1 is A_1 and x_2 is A_2 and … and x_n is A_n, then y is B

The presidency part may contain any number of "and/or"

Zadeh [6] fuzzy inference is given as:

If x_1 is A_1 and x_2 is A_2 and … and x_n is A_n, then y is B

$= \min(1, 1 - (A_1, A_2, \ldots, A_n) + B)$

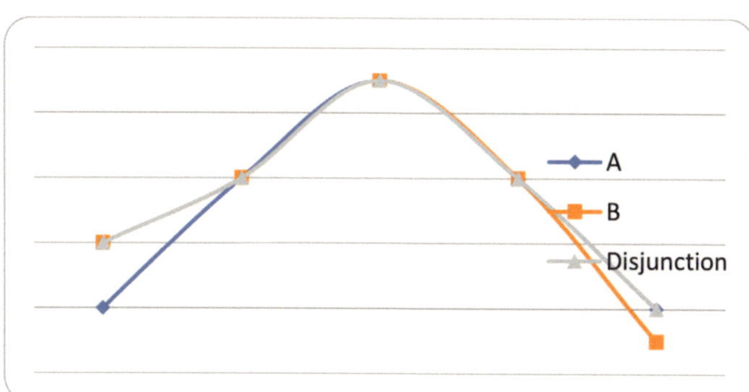

Figure 3.
Disjunction.

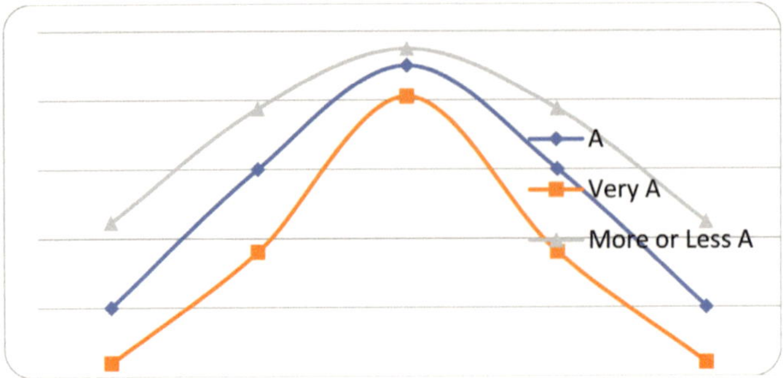

Figure 4.
Fuzzy quantifiers.

Mamdani [4] fuzzy inference is given as:

If x_1 is A_1 and x_2 is A_2 and … and x_n is A_n, then y is B

$= \min(A_1, A_2,\ldots, A_n, B)$

Zadeh and Mamdani fuzzy inference has prior information of A and B. The relation between A and B is known. Then, B is derived from A.

Reddy [2] inference is given by:

If x_1 is A_1 and x_2 is A_2 and … and x_n is A_n, then y is B

$= \min(A_1, A_2,\ldots,A_n)$

Consider the fuzzy rule:

If x_1 is A_1 and x_2 is A_2, then x is B

For instance,

$A_1 = 0.2/x_1 + 0.6/x_2 + 0.9/x_3 + 0.6/x_4 + 0.2/x_5$

$A_2 = 0.5/x_1 + 0.7/x_2 + 0.9/x_3 + 0.7/x_4 + 0.3/x_5$

$B = 0.1/x_1 + 0.4/x_2 + 0.6/x_3 + 0.4/x_4 + 0.1/x_5$

The graphical representation of A1, A2 and B is shown in **Figure 5**.

The graphical representation of fuzzy inference is shown in **Figure 6**.

Composition

If some relation between R and A1 than B1 is to infer from R

B1 = A1 o R, where R = A→B

Zadeh fuzzy inference is given by:

$B1 = A1 \text{ o } R = \min\{\mu_A(x), \mu_R(x)\}$

$= \min\{\mu_A(x), \min(1,1-\mu_{A1}(x) + \mu_B(x))\}$

Mamdani fuzzy inference is given by:

$= \min\{\mu_{A1}(x),\mu_A(x) + \mu_B(x)\}$

If there is some relation R between A and B, then Reddy fuzzy inference is given by:

$= \mu_{A1}(x)$

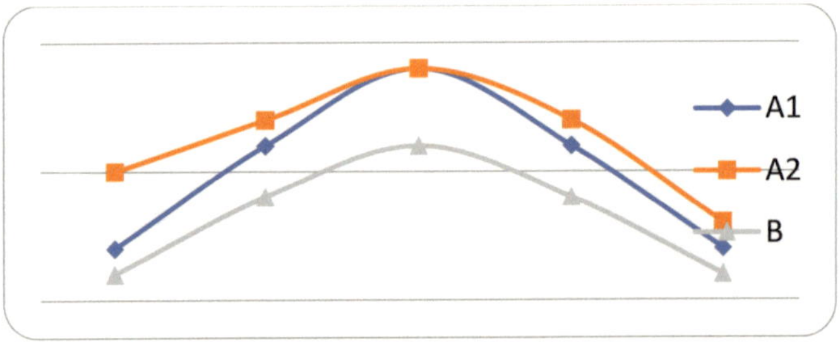

Figure 5.
Fuzzy sets.

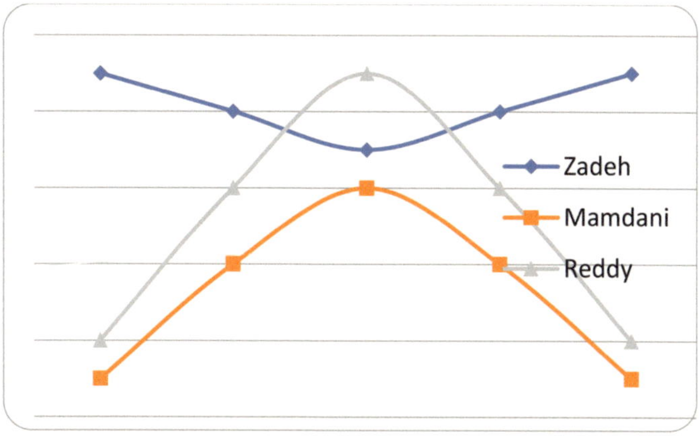

Figure 6.
Fuzzy conditional inference.

3. Justification of Reddy and Mamdani fuzzy conditional inference

Justification of Reddy fuzzy conditional inference may be derived in the following:

Consider Reddy fuzzy conditional inference:

If x_1 is A_1 and x_2 is A_2 and … and x_n is A_n, then y is B = min{A_1, A_2,…, A_n}.

Consider TSK fuzzy conditional inference:

If x_1 is A_1 and x_2 is A_2 and … and x_n is A_n, then y is B = f(x_1, x_2,…, x_n).

The proposed method of fuzzy conditional inference may be defined by replacing x_1, x_2,…, x_n with A_1, A_2 and … and A_n

If x_1 is A_1 and/or A_2 and/or,…, and/or A_n, then y is B = f(A_1, A_2,…, A_n)

If x_1 is A_1 or A_2 and A_n, then y is B = f(A_1, A_2, A_3) = A_1 V A_2 Λ –Λ A_3

If x_1 is A_1 or A_2 and A_3, then y is B = f(A_1, A_2, A_3) = A_1 V A_2 Λ A_3

B = min(max($\mu_{A1}(x_1)$, $\mu_{A2}(x_2)$), $\mu_{A3}(x_3)$)

The fuzzy conditional inference is given by using Mamdani fuzzy inference

If x_1 is A_1 or A_2 and A_3, then y is B = min(A1 or A2 and A3, B)

If x_1 is A_1 or A_2 and A_3, then x is B = min(max($\mu_{A1}(x_1)$, $\mu_{A2}(x_2)$), $\mu_{A3}(x_3)$)

Thus, the Reddy fuzzy conditional inference is satisfied.

If x_1 is A_1 and x_2 is A_2 and … and x_n is A_n, then y is B = min{A_1, A_2,…,A_n}.

Justification of Mamdani fuzzy conditional inference may be derived in the following:

If some relation R between A and B is known, then Mamdani fuzzy conditional inference is given by:

If x is A, then y is B = A x B

Zadeh fuzzy conditional inference for "if … then … else …" is given by:

If x is A, then y is B else y is C = A x B v A′ x C

If x is A, then y is B else y is C = If x is A then y is B v If x is A′ then y is C = A x B v A′ x C

It is logically divided into:

If x is A, then y is B = A x B

If x is A′, then y is C = A′ x C

Thus, the Mamdani fuzzy conditional inference is satisfied.

If x is A, then y is B = A x B.

4. Fuzzy control systems using single fuzzy membership function

Zadeh introduced fuzzy algorithms. The fuzzy algorithm is a set of fuzzy statements. The fuzzy conditional statement is defined as fuzzy algorithm:

If x_i is $A1_i$ and x_i is $A2_i$ and… and x_i is An, then y is B_i

The consequent part may not be known in control systems

The fuzziness may be given for Reddy fuzzy inference as

If BZ is low (0.6)

and BE is normal (0.7)

then reduce fan speed

= min (0.6, 0.7)

= 0.6

The fuzzy set type-2 is a type of fuzzy set in which some additional degree of information is provided.

Definition: Given some universe of discourse X, a fuzzy set type-2 A of X is defined by its membership function $\mu_A(x)$ taking values on the unit interval [0,1], i.e., $\mu_{\tilde{A}}(x) \rightarrow [0,1]^{[0.1]}$

Suppose X is a finite set. The fuzzy set A of X may be represented as

$A = \mu_{\tilde{A}1}(x_1)/\tilde{A}_1 + \mu_{\tilde{A}2}(x_2)/\tilde{A}_2 + \ldots + \mu_{\tilde{A}n}(x_n)/\tilde{A}n$

Temperature = {0.4/low, 0.6/medium, 0.9/high}

John has "mild headache" with fuzziness 0.4

The fuzzy control system for boiler consists of a set of fuzzy rules [4].

If a set of conditions is satisfied, then the set of consequences is fired

The fuzzy control system is shown in **Figure** 7.

The fuzzy control system containing fuzzy variables are represented in decision **Table 1**.

The fuzzy control system of boiler is given in **Table** 2.

For instance,

If BZ is low

and BE is normal

then reduce fan speedFor instance, consider the fuzzy control system (**Table 3**).

The computation of proposed method (3.4) is given in **Table 4**.

Defuzzification

The centroid technique is used for defuzzification. It finds value representing the centre of gravity (COG) aggregated fuzzy generalized fuzzy set:

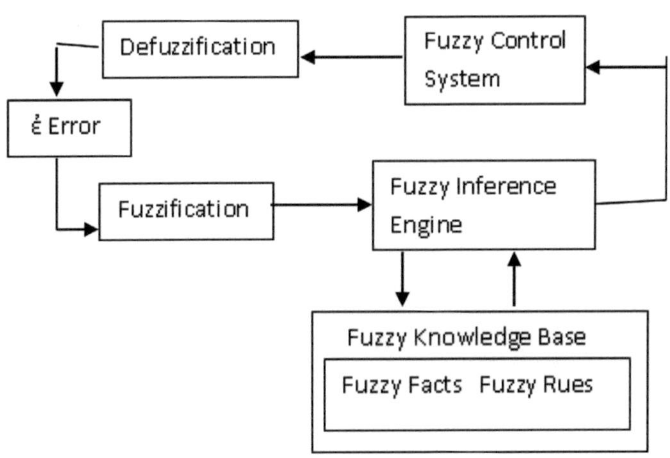

Figure 7.
Fuzzy control system.

A1	A2	…	An	B
A11	A12	…	A1n	B1
A21	A22	…	A2n	B2
⋮	⋮		⋮	⋮
Am1	Am2	…	Amn	Bmn

Table 1.
Fuzzy rules.

Condition	Burning zone (BZ) temperature	Back-end (BE) temperature	Action
AND	Drastically low	Low	Reduce Klin speed
AND	Drastically low	Low	Reduce fuel
AND	Slightly low	Low	Increase fan speed
AND	Low	High	Reduce fuel
AND	Low	Normal	Reduce fan speed

Table 2.
Boiler controller.

Condition	Burning zone (BZ) temperature	Back-end (BE) temperature	Action
AND	Drastically low (0.7)	Low (0.6)	Reduce Klin speed
AND	Drastically low (0.7)	Low (0.8)	Reduce fuel
AND	Slightly low (.8)	Low (.9)	Increase fan speed
AND	Low (0.7)	High (0.65)	Reduce fuel
AND	Low (0.6)	Normal (0.7)	Reduce fan speed

Table 3.
Boiler fuzzy controller.

Condition	Burning zone (BZ) temperature	Back-end (BE) temperature	Action
AND	Drastically low (0.7)	Low (0.6)	Reduce Klin speed (0.6)
AND	Drastically low (0.7)	Low (0.8)	Reduce fuel (0.7)
AND	Slightly low (.8)	Low (.9)	Increase fan speed (0.8)
AND	Low (0.7)	High (0.65)	Reduce fuel (0.65)
AND	Low (0.6)	Normal (0.7)	Reduce fan speed (0.6)

Table 4.
Fuzzy inference.

$COG = \Sigma C_i \mu_{Ai}(x) / \Sigma C_i$
For instance,
$Speed = \{0.1/20 + 0.3/40 + 0.5/60 + 0.7/80 + 0.9/100\}$
$COG = (0.1{*}20 + 0.3{*}40 + 0.5{*}60 + 0.7{*}80 + 0.9{*}100)/$
$(0.1 + 0.3 + 0.5 + 0.7 + 0.9) = 73.6$

Condition	Burning zone (BZ) temperature	Back-end (BE) temperature	Action
AND/OR	Drastically low (0.7,0.1)	Low (0.8,0.1)	Reduce Klin speed (0.6,0.2)
AND/OR	Drastically low (0.8,0.1)	Low (0.9,0.1)	Reduce fuel (0.7,0.2)
AND/OR	Slightly low (1.0,0.2)	Low (1.0,0.1)	Increase fan speed (0.9,0.2)
AND/OR	Low (0.8,0.1)	High (0.9,0.2)	Reduce fuel (0.6,0.1)
AND/OR	Low (0.7,0.1)	Normal (0.8,0.2)	Reduce fan speed (0.5,0.1)

Table 5.
Twofold fuzziness.

5. Fuzzy logic with twofold fuzzy sets

Generalized fuzzy logic is studied for incomplete information [8, 9].

Given some universe of discourse X, the proposition "x is A" is defined as its twofold fuzzy set with membership function as

$\mu_A(x) = \{\mu_A^{True}(x), \mu_A^{False}(x)\}$

or

$A = \{\mu_A^{True}(x), \mu_A^{False}(x)\}$

where A is the seneralized fuzzy set and x ∈ X,

$0 <= \mu_A^{True}(x) <= 1$ and, $0 <= \mu_A^{False}(x) <= 1$

$A = \{\mu_A^{True}(x_1)/x_1 + \ldots + \mu_A^{True}(x_n)/x_n,$

$\mu_A^{False}(x_1)/x_1 + \ldots + \mu_A^{True}(x_n)/x_n, x_i \in X,$

$\mu_A^{True}(x) + \mu_A^{False}(x) < 1,$

$\mu_A^{True}(x) + \mu_A^{False}(x) > 1$ and

$\mu_A^{True}(x) + \mu_A^{False}(x) = 1$

The conditions are interpreted as redundant, insufficient and sufficient, respectively.

For instance,

$A = \{0.5/x_1 + 0.7/x_2 + 0.9/x_3 + 0.7/x_4 + 0.5/x_5,$

$0.1/x_1 + 0.2/x_2 + 0.3/x_3 + 0.2/x_4 + 0.1/x_5\}$

The graphical representation is shown in **Figure 8**.

The fuzzy logic is defined as a combination of fuzzy sets using logical operators. Some of the logical operations are given below.

Let A, B and C be the fuzzy sets. The operations on fuzzy sets are given below for twofold fuzzy sets.

Negation

$A' = \{1-\mu_A^{True}(x), 1-\mu_A^{False}(x)\}/x$

Disjunction

$AVB = \{\max(\mu_A^{True}(x), \mu_A^{True}(y)), \max(\mu_B^{False}(x), \mu_B^{False}(y))\}(x,y)$

Conjunction

$A\Lambda B = \{\min(\mu_A^{True}(x), \mu_A^{True}(y)), \min(\mu_B^{False}(x), \mu_B^{False}(y))\}/(x,y)$

Composition

$A \text{ o } R = \{\min_x (\mu_A^{True}(x), \mu_A^{True}(x)), \min_x(\mu_R^{False}(x), \mu_R^{False}(x))\}/y$

The fuzzy propositions may contain quantifiers like "very", "more or less". These fuzzy quantifiers may be eliminated as follows:

Concentration

"x is very A"

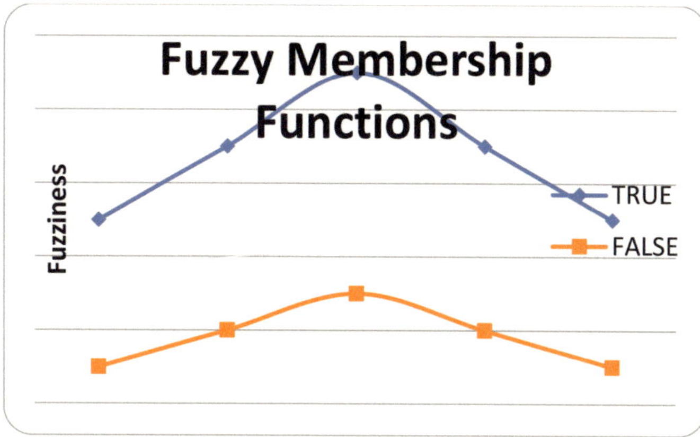

Figure 8.
Fuzzy membership function.

$\mu_{very\ A}(x) = \{\mu_A^{True}(x)^2, \mu_A^{False}(x)\mu_A(x)^2\}$

Diffusion

"x is more or less A"

$\mu_{more\ or\ less\ A}(x) = (\mu_A^{True}(x)^{1/2}, \mu_A^{False}(x)\mu_A(x)^{0.5}$

$A = \{0.5/x_1 + 0.7/x_2 + 0.9/x_3 + 0.7/x_4 + 0.5/x_5,$
$0.1/x_1 + 0.2/x_2 + 0.3/x_3 + 0.2/x_4 + 0.1/x_5\}$

$B = \{0.4/x_1 + 0.6/x_2 + 0.8/x_3 + 0.6/x_4 + 0.4/x_5,$
$0.1/x_1 + 0.2/x_2 + 0.3/x_3 + 0.2/x_4 + 0.1/x_5\}$

$A' = \text{not } A = \{0.5/x_1 + 0.3/x_2 + 0.1/x_3 + 0.3/x_4 + 0.5/x_5,$
$0.9/x_1 + 0.8/x_2 + 0.7/x_3 + 0.8/x_4 + 0.9/x_5\}$

$A \vee B = \{0.5/x_1 + 0.7/x_2 + 0.9/x_3 + 0.7/x_4 + 0.5/x_5,$
$0.1/x_1 + 0.2/x_2 + 0.3/x_3 + 0.2/x_4 + 0.1/x_5\}$

$A \wedge B = \{0.4/x_1 + 0.6/x_2 + 0.8/x_3 + 0.6/x_4 + 0.4/x_5,$
$0.1/x_1 + 0.2/x_2 + 0.3/x_3 + 0.2/x_4 + 0.1/x_5\}$

$\text{Very } A = \{0.25/x_1 + 0.49/x_2 + 0.81/x_3 + 0.49/x_4 + 0.25/x_5,$
$0.01/x_1 + 0.04/x_2 + 0.09/x_3 + 0.04/x_4 + 0.01/x_5\}$

$\text{More or less } A = \{0.70/x_1 + 0.83/x_2 + 0.94/x_3 + 0.83/x_4 + 0.70/x_5,$
$0.31/x_1 + 0.44/x_2 + 0.54/x_3 + 0.44/x_4 + 0.31/x_5\}$

$A \rightarrow B = \{1/x_1 + 0.8/x_2 + /x_3 + 0.9/x_4 + 1/x_5,$
$1/x_1 + 1/x_2 + 1/x_3 + 0.8/x_4 + 1/x_5\}$

$A \circ B = \{0.8/x_1 + 0.7/x_2 + 0.7/x_3 + 0.5/x_4 + 0.5/x_5,$
$0.4/x_1 + 0.3/x_2 + 0.4/x_3 + 0.5/x_4 + 0.6/x_5\}$

Implication

Consider the fuzzy condition "if x is A_1 and x is A_2 and .. and x is A_n, then y is B."

The presidency part may contain any number of "and"/"or."

Zadeh fuzzy conditional inference given as

$= \{\min(1, 1 - \min(\mu_{A1}^{True}(x), \mu_{A2}^{True}(x), \ldots, \mu_{An}^{True}(x)) + \mu_B^{True}(y)),$
$\min(1, 1 - \min(\mu_{A1}^{False}(x), \mu_{A2}^{TrueFalse}(x), \ldots, \mu_{An}^{False}(x)) + \mu_B^{False}(y))\}(x,y)$

Mamdani fuzzy conditional inference given as

$= \{\min(\mu_{A1}^{True}(x), \mu_{A2}^{True}(x), \ldots, \mu_{An}^{True}(x), \mu_B^{True}(y)), \min(\mu_{A1}^{False}(x),$
$\mu_{A2}^{TrueFalse}(x), \ldots, \mu_{An}^{False}(x), \mu_B^{False}(y))\}(x,y)$

Reddy [5] fuzzy conditional inference given by

$= \{\min(\mu_{A1}^{True}(x), \mu_{A2}^{True}(x), \ldots, \mu_{An}^{True}(x)), \min(\mu_{A1}^{False}(x), \mu_{A2}^{TrueFalse}(x), \ldots,$
$\mu_{An}^{False}(x))\}(x,y)$

Consider the fuzzy condition "if x is A_1 and x is A_2, then x is B"

The presidency part may contain any number of "and"/"or."

For instance,

$A1 = \{0.5/x_1 + 0.7/x_2 + 0.9/x_3 + 0.7/x_4 + 0.5/x_5,$
$0.1/x_1 + 0.2/x_2 + 0.3/x_3 + 0.2/x_4 + 0.1/x_5\}$

$A2 = \{0.4/x_1 + 0.6/x_2 + 0.8/x_3 + 0.6/x_4 + 0.4/x_5,$
$0.1/x_1 + 0.2/x_2 + 0.3/x_3 + 0.2/x_4 + 0.1/x_5\}$

$B = \{0.5/x_1 + 0.7/x_2 + 1/x_3 + 0.7/x_4 + 0.5/x_5,$
$0.4/x_1 + 0.5/x_2 + 0.6/x_3 + 0.5/x_4 + 0.4/x_5\}$

Zadeh fuzzy conditional inference given as

$=\{\min(1, 1-\min(\mu_{A1}^{True}(x), \mu_{A2}^{True}(x)) + \mu_B^{True}(x)), \min(1, 1-\min(\mu_{A1}^{False}(x), \mu_{A2}^{TrueFalse}(x)) + \mu_B^{False}(x))\}$

$= \{1/x_1 + 0.1/x_2 + 1/x_3 + 1/x_4 + 1/x_5,$
$1/x_1 + 1/x_2 + 1/x_3 + 1/x_4 + 1/x_5\}$

Mamdani fuzzy conditional inference given as

$= \{\min(\mu_{A1}^{True}(x), \mu_{A2}^{True}(x), \ldots, \mu_{An}^{True}(x), \mu_B^{True}(x)), \min(\mu_{A1}^{False}(x), \mu_{A2}^{TrueFalse}(x), \ldots, \mu_{An}^{False}(x), \mu_B^{False}(x))\}$

$= \{0.4/x_1 + 0.6/x_2 + 0.8/x_3 + 0.6/x_4 + 0.4/x_5,$
$0.1/x_1 + 0.2/x_2 + 0.3/x_3 + 0.2/x_4 + 0.1/x_5\}$

Reddy fuzzy conditional inference given as

$=\{\min(\mu_{A1}^{True}(x), \mu_{A2}^{True}(x)), \min(\mu_{A1}^{False}(x), \mu_{A2}^{TrueFalse}(x))\}$

$= \{0.4/x_1 + 0.6/x_2 + 0.8/x_3 + 0.6/x_4 + 0.4/x_5,$
$0.1/x_1 + 0.2/x_2 + 0.3/x_3 + 0.2/x_4 + 0.1/x_5\}$

Composition

If some relation R between A and B is known and some value A1 than B1 is inferred from R,

$B1 = A1 \circ R,$

where $R = A \rightarrow B$

Zadeh fuzzy inference is given by

$B1 = A1 \circ R == A1 \circ \{\min(1, 1 - \mu_A^{True}(x) + \mu_B^{True}(x)), \min(1, 1 - \mu_A^{False}(x) + \mu_B^{False}(x))\}$

$= \min\{\mu_A(x), \min(1, 1- \mu_{A1}(x) + \mu_B(x)\}$

Mamdani fuzzy inference is given by

$= A1 \circ \{\min(\mu_A^{True}(x), \mu_B^{True}(x)), \min(\mu_A^{TrueFalse}(x), \mu_B^{False}(x)\}$

If some relation R between A and B is not known,

according to Reddy fuzzy inference,

$= \{\min(\mu_{A1}^{True}(x), \mu_A^{True}(x)), \min(\mu_{A1}^{TrueFalse}(x), \mu_A^{False}(x)\}$

The fuzzy set A of X is characterized as its membership function $A = \mu_A(x)$ and ranging values in the unit interval [0, 1]

$\mu_A(x): X \rightarrow [0, 1], x \in X$, where X is universe of discourse.

$A = \mu_A(x) = \mu_A(x_1)/x_1 + \mu_A(x_2)/x_2 + \ldots + \mu_A(x_n)/x_n$, "+" is union

The generalized fuzzy certainty factor (GFCF) is defined as

$\mu_A^{GFCF}(x) = \mu_A^{True}(x) - \mu_A^{False}(x)$

The generalized fuzzy certainty factor becomes single fuzzy membership function.

$\mu_A^{GFCF(x)}: X \rightarrow [0, 1], x \in X$, where X is universe of discourse.

The generalized fuzzy certainty factor (GFCF) will compute the conflict of evidence in the uncertain information.

For example,

$A = \{0.5/x_1 + 0.7/x_2 + 0.9/x_3 + 0.7/x_4 + 0.5/x_5,$
$0.1/x_1 + 0.2/x_2 + 0.3/x_3 + 0.2/x_4 + 0.1/x_5\}$

$\mu_A^{GFCF}(x) = \{0.5/x_1 + 0.7/x_2 + 0.9/x_3 + 0.7/x_4 + 0.5/x_5 - 0.1/x_1 + 0.2/x_2 + 0.3/x_3 + 0.2/x_4 + 0.1/x_5\}$

$= 0.4/x_1 + 0.5/x_2 + 0.6/x_3 + 0.5/x_4 + 0.4/x_5$

For instance, "x is high temperature" with fuzziness {0.8,0.2}

The GFCF is 0.6
The graphical representation of GFCF is shown in **Figure 9**.
For example, A and B are generalized fuzzy sets.
$A = \{0.5/x_1 + 0.7/x_2 + 0.9/x_3 + 0.7/x_4 + 0.5/x_5 -$
$0.1/x_1 + 0.2/x_2 + 0.3/x_3 + 0.2/x_4 + 0.1/x_5\}$
$= 0.4/x_1 + 0.5/x_2 + 0.6/x_3 + 0.5/x_4 + 0.4/x_5$
$B = \{0.4/x_1 + 0.6/x_2 + 0.8/x_3 + 0.6/x_4 + 0.4/x_5 =$
$0.1/x_1 + 0.2/x_2 + 0.3/x_3 + 0.2/x_4 + 0.1/x_5\}$
$= 0.3/x_1 + 0.4x_2 + 0.5/x_3 + 0.4/x_4 + 0.3/x_5$
The operations on GFCF are given as follows:
Negation
$A' = 1 - \mu_A{}^{GFCF}(x)/x$
$= 0.6/x_1 + 0.5/x_2 + 0.4/x_3 + 0.5x_4 + 0.6/x_5$
The graphical representation is shown in **Figure 10**.
Conjunction
$A \wedge B = \min(\mu_A(x), \mu_B(x)\}/x$
$A \wedge B = 0.3/x_1 + 0.4/x_2 + 0.5/x_3 + 0.4/x_4 + 0.3/x_5$
The graphical representation is shown in **Figure 11**.
Disjunction
$AVB = \max(\mu_A(x), \mu_B(y)\}/x$
$AVB = .4/x_1 + 0.6/x_2 + 0.9/x_3 + 0.6/x_4 + 0.2/x_5$
The graphical representation is shown in **Figure 12**.
Concentration
$\mu_{vey\ A}{}^{GFCF}(x) = \mu_A{}^{GFCF}(x)^2$

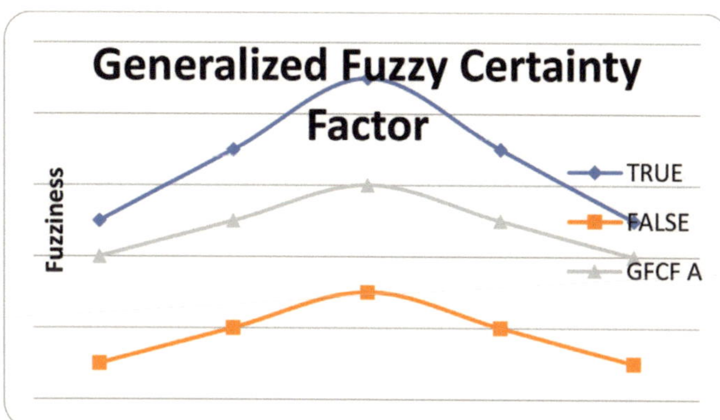

Figure 9.
Generalized fuzzy certainty factor.

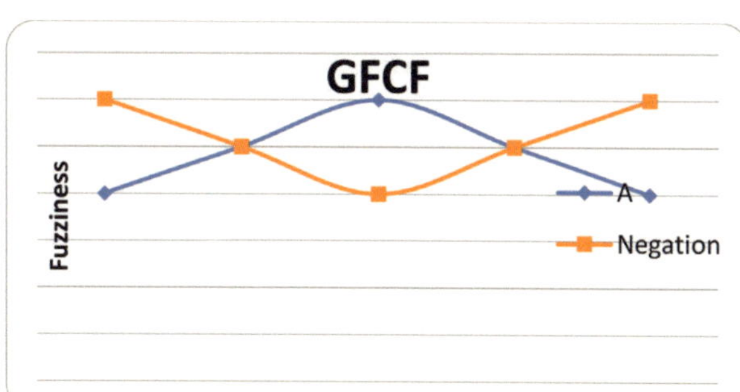

Figure 10.
Negation.

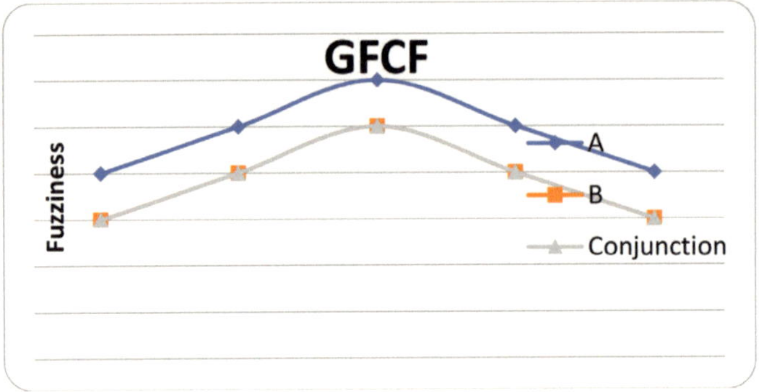

Figure 11.
Conjunction.

$= 0.16/x_1 + 0.25/x_2 + 0.36/x_3 + 0.25/x_4 + 0.16/x_5$

Diffusion

$\mu_{\text{more or less A}}{}^{GFCF}(x) = \mu_A{}^{GFCF}(x)^{0.5}$

$= 0.63/x_1 + 0.71/x_2 + 0.77/x_3 + 0.71/x_4 + 0.63/x_5$

The graphical representation of concentration and diffusion are shown in **Figure 13**.

Implication

Zadeh [9], Mamdani [7] and Reddy [5] fuzzy conditional inferences are considered keeping in view of fuzzy control systems.

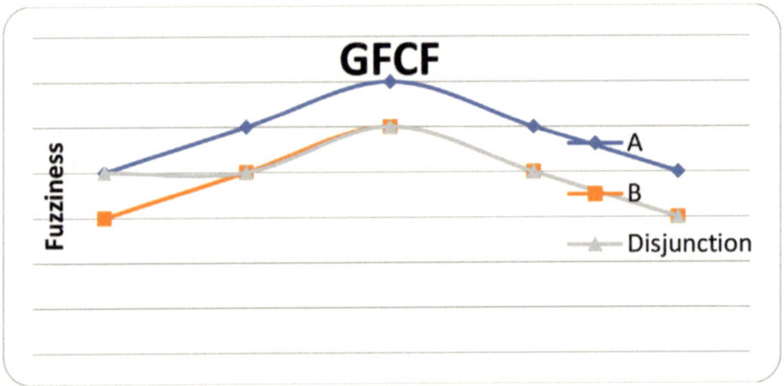

Figure 12.
Disjunction.

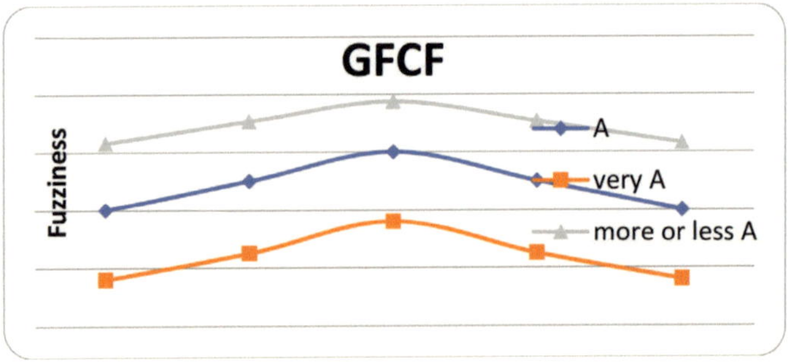

Figure 13.
Implication.

If x_1 is A_1 and x_2 is A_2 and … and x_n is A_n, then y is B
The presidency part may contain any number of "and"/"or."
Zadeh fuzzy inference is given as follows:
If x_1 is A_1 and x_2 is A_2 and … and x_n is A_n, then y is B
$= \min(1, 1 - (A_1, A_2, \ldots, A_n) + B)$
Mamdani fuzzy inference is given as follows:
If x_1 is A_1 and x_2 is A_2 and … and x_n is A_n, then y is B
$= \min(A_1, A_2, \ldots, A_n, B)$
Reddy inference is given as follows:
If x_1 is A_1 and x_2 is A_2 and … and x_n is A_n, then y is B
$= \min(A_1, A_2, \ldots, A_n)$
Consider the fuzzy rule:
If x_1 is A_1 and x_2 is A_2, then x is B
For instance,
$A1 = \{0.5/x_1 + 0.7/x_2 + 0.9/x_3 + 0.7/x_4 + 0.5/x_5 -$
$0.1/x_1 + 0.2/x_2 + 0.3/x_3 + 0.2/x_4 + 0.1/x_5\}$
$= 0.4/x_1 + 0.5/x_2 + 0.6/x_3 + 0.5/x_4 + 0.4/x_5$
$A2 = \{0.4/x_1 + 0.6/x_2 + 0.8/x_3 + 0.6/x_4 + 0.4/x_5 =$
$0.1/x_1 + 0.2/x_2 + 0.3/x_3 + 0.2/x_4 + 0.1/x_5\}$
$= 0.3/x_1 + 0.4x_2 + 0.5/x_3 + 0.4/x_4 + 0.3/x_5$
$B = \{0.5/x_1 + 0.7/x_2 + 1/x_3 + 0.7/x_4 + 0.5/x_5,$
$0.4/x_1 + 0.5/x_2 + 0.6/x_3 + 0.5/x_4 + 0.4/x_5\}$
$= 0.1/x_1 + 0.2x_2 + 0.4/x_3 + 0.2/x_4 + 0.1/x_5$
The graphical representation of A1, A2 and B is shown in **Figure 14**.
Zadeh fuzzy inference is given as
$= \min(1, 1 - (A_1, A_2) + B)$
$= 0.8/x_1 + 0.8/x_2 + 0.9/x_3 + 0.8/x_4 + 0.8/x_5$
Mamdani fuzzy inference is given as
$\min(A_1, A_2, \ldots, A_n, B)$
$= 0.1/x_1 + 0.2/x_2 + 0.4/x_3 + 0.2/x_4 + 0.1/x_5$
Reddy fuzzy inference is given as
$\min(A_1, A_2, \ldots, A_n)$
$= 0.2/x_1 + 0.4/x_2 + 0.5/x_3 + 0.4/x_4 + 0.3/x_5$
The graphical representation of fuzzy inference is shown in **Figure 15**.
Composition
The GFCF is a single fuzzy membership function

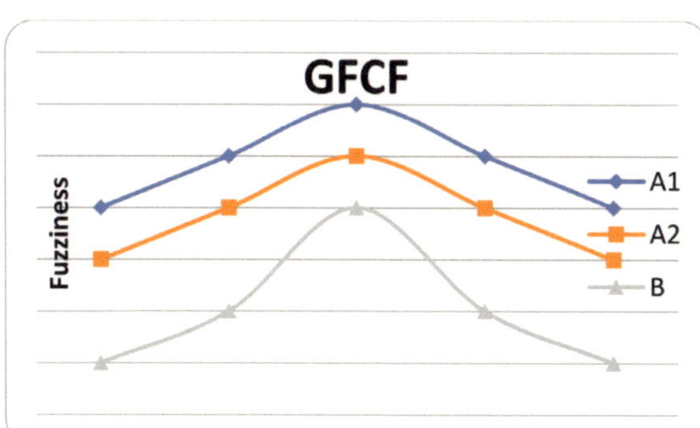

Figure 14.
GFCF for fuzzy rule.

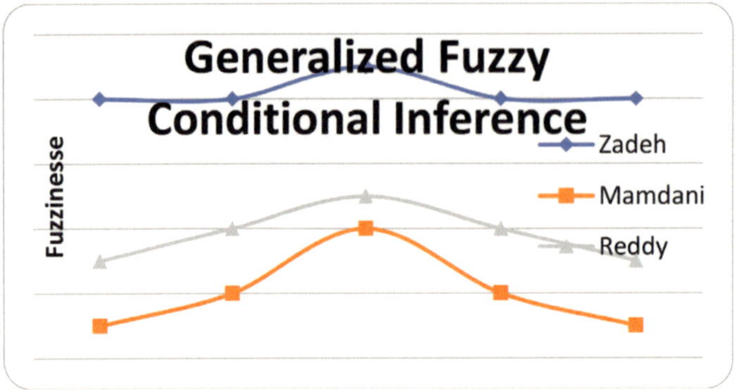

Figure 15.
Implication.

If some relation R between A1, then B1 is to infer from R:
$B1 = A1 \text{ o } R = \min\{\mu_{A1}{}^{GFCF}(x), \mu_R{}^{GFCF}(x)\}/x$
Zadeh fuzzy inference is given by
$B1 = A1 \text{ o } R = \min\{\mu_{A1}{}^{GFCF}(x), \mu_R{}^{GFCF}(x)\}$
$= \min\{\mu_{A1}{}^{GFCF}(x), \min(1, 1 - \mu_{A1}{}^{GFCF}(x) + \mu_B{}^{GFCF}(x))\}$
Mamdani fuzzy inference is given by
$= \min\{\mu_{A1}{}^{GFCF}(x), \mu_{A1}{}^{GFCF}(x), \mu_B{}^{GFCF}(x)\}$
If there is some relation R between A and B, then Reddy fuzzy inference is given by
$= \mu_{A1}{}^{GFCF}(x)$
where A, B, A1, and B1 are the GFCF.
$A = \{0.5/x_1 + 0.7/x_2 + 0.9/x_3 + 0.7/x_4 + 0.5/x_5 -$
$0.1/x_1 + 0.2/x_2 + 0.3/x_3 + 0.2/x_4 + 0.1/x_5\}$
$= 0.4/x_1 + 0.5/x_2 + 0.6/x_3 + 0.5/x_4 + 0.4/x_5$
$B = \{0.5/x_1 + 0.7/x_2 + 1/x_3 + 0.7/x_4 + 0.5/x_5,$
$0.4/x_1 + 0.5/x_2 + 0.6/x_3 + 0.5/x_4 + 0.4/x_5\}$
$= 0.1/x_1 + 0.2x_2 + 0.4/x_3 + 0.2/x_4 + 0.1/x_5$
A1 = more or less A
$= = 0.55/x_1 + 0.63/x_2 + 0.71/x_3 + 0.63/x_4 + 0.55/x_5$
The composition of Zadeh, Mamdani and Reddy fuzzy inference is shown in **Figure 16**.

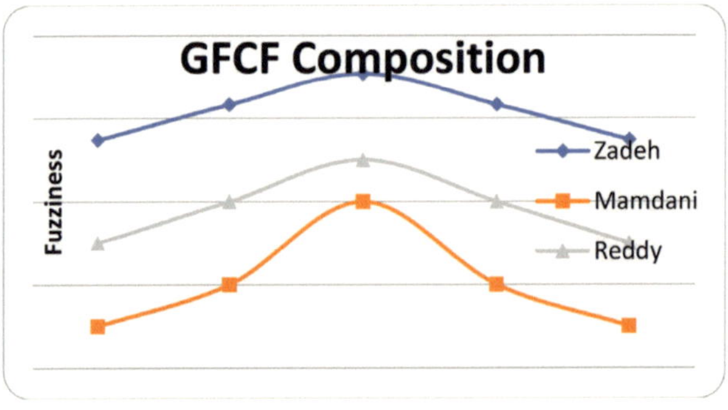

Figure 16.
Composition.

6. Fuzzy control systems using two fuzzy membership functions

Zadeh [6] introduced fuzzy algorithms. The fuzzy algorithm is a set of fuzzy statements. The fuzzy conditional statement is defined as follows:

If x_i is $A1_i$ and x_i is $A2_i$ and … and x_i is An, then y_i is B_i

The precedence part may contain and/or/not.

The fuzzy control system consist of a set of fuzzy rules.

If a set of conditions is satisfied, then a set of consequences is inferred.

The fuzzy set with twofold membership function will give more information than the single membership function.

The generalized fuzzy certainty factor (GFCF) is given as

$\mu_A{}^{GFCF}(x) = \{\mu_A{}^{True}(x) - \mu_A{}^{False}(x)\}$

For instance, "x has fever"

The GFCF for fever given as

$\mu_{Low}{}^{GFCF}(x) = \{\mu_{Low}{}^{True}(x) - \mu_{Low}{}^{False}(x)\}$

Consider the rule in fuzzy control system

If BZ is low

and BE is normal

then reduce fan speed

For instance, fuzziness may be given as follows:

If BZ is low (0.9,0.2)

and BE is normal (0.8,0.2)

then reduce fan speed (0.6, 0.3)

Fuzziness of GFCF may be given as follows:

If BZ is low (0.7)

and BE is normal (0.6)

then reduce fan speed (0.3)

For instance, consider the twofold fuzzy relational model of fuzzy control system.

The graphical representation of twofold fuzzy relational model is shown in **Figure 17**.

The graphical representation of fuzzy inference for condition part containing "AND" is shown in **Figure 18**.

Graphical representation of fuzzy inference for condition part containing "OR" is shown in **Figure 19**.

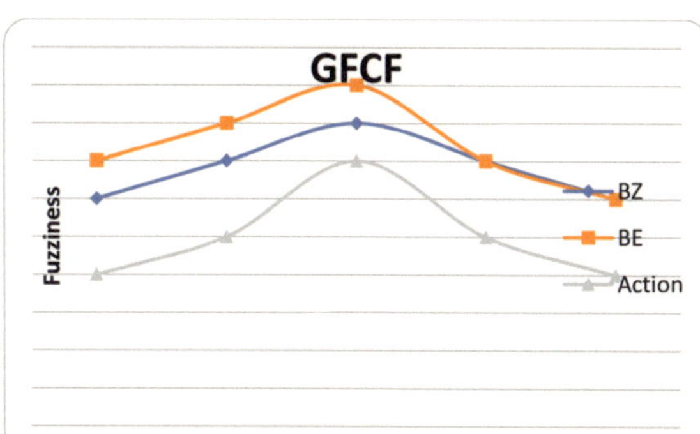

Figure 17.
*GFCF for **Table 5**.*

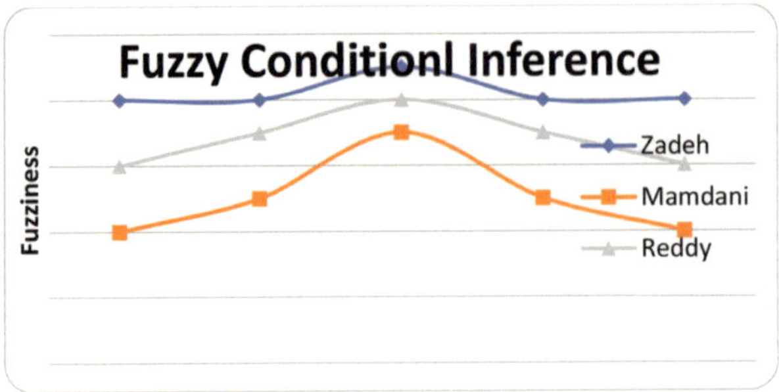

Figure 18.
Fuzzy conditional inference for "AND."

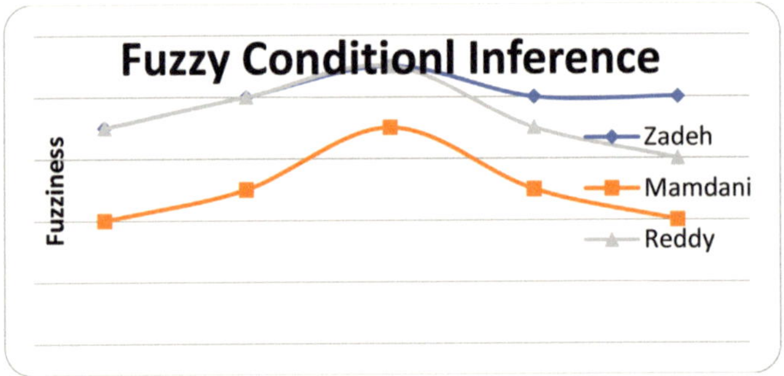

Figure 19.
Fuzzy conditional inference for "OR."

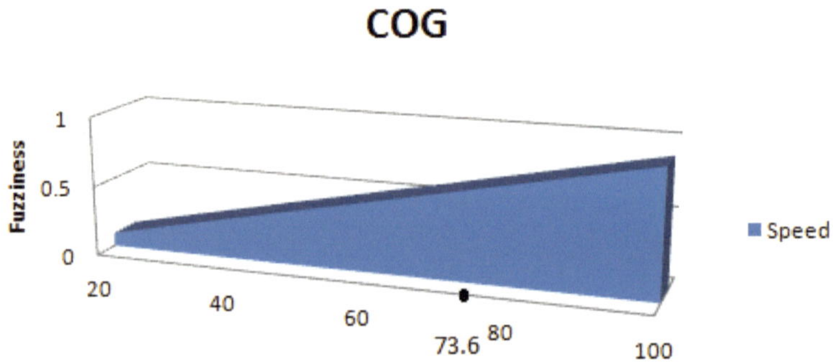

Figure 20.
Defuzzification.

Defuzzification

Usually, centroid technique is used for defuzzification. It finds value representing the centre of gravity (COG) aggregated fuzzy generalized fuzzy set.

$COG = \Sigma\, C_i\, \mu_{Ai}{}^{GFCF}(x) / \Sigma\, C_i$

For instance,

Speed = {0.1/20 + 0.3/40 + 0.5/60 + 0.7/80 + 0.9/100}

COG = (0.1*20 + 0.3*40 + 0.5*60 + 0.7*80 + 0.9*100)/
(0.1 + 0.3 + 0.5 + 0.7 + 0.9) = 73.6

The defuzzification is shown in **Figure 20**.

7. Conclusion

The fuzzy set of two membership function will give more information than single fuzzy membership function for incomplete information. The fuzzy logic and fuzzy conditional inference based on single membership function and twofold fuzzy set are studied. The FCF is studied as difference between "True" and "False" membership functions to eliminate conflict of evidence and to make as single fuzzy membership function. FCF = [True-False] will correct truthiness of single membership function. The methods of Zadeh, Mamdani and Reddy fuzzy conditional inference studied for fuzzy control systems are given as application.

Conflict of interest

The author states that he has no conflict of interest and that he has permission to use parts of his previously published work from the original publisher.

Author details

Poli Venkata Subba Reddy
Department of Computer Science and Engineering, College of Engineering,
Sri Venkateswara University, Tirupati, India

*Address all correspondence to: pvsreddy@hotmail.co.in

References

[1] Ping REN. Generalized fuzzy sets and representation of incomplete knowledge. Fuzzy Sets and Systems. 1990;**1**(36):91-96

[2] Takagi T, Sugen M. Fuzzy identification of systems and its application to modeling and control. IEEE Transactions on Systems, Man, and Cybernetics. 1985;**15**(1):116-132

[3] Reddy PVS. Generalized fuzzy logic for incomplte information. In: IEEE International Conference on Fuzzy Systems, Proceedings (IEEE-FUZZ 2013); Hyderabad, India; July 7–10, 2013

[4] Mamdani EH, Assilian S. An experiment in linguistic synthesis with a fuzzy logic controller. International Journal of Man-Machine Studies. 1975;7: 1-13

[5] Reddy PVS. Fuzzy logic based on belief and disbelief membership functions. Fuzzy Information and Engineering. 2018;**9**(4):405-422

[6] Zadeh LA. Fuzzy sets. Information and Control. 1965;**8**(3):38-353

[7] Reddy PVS. Some methods on fuzzy conditional inference to approximate reasoning. In: 2016 International Conference on Fuzzy Theory and Its Applications (iFUZZY 2016); IEEE XPlore; November 9–11, 2016; Taichung, Taiwania

[8] Zadeh LA. Generalized theory of uncertainty (GTU). Principal concepts and ideas. Computational Statistics & Data Analysis. 2006;**51**:15-46

[9] Zadeh LA. The role of fuzzy logic in the management if uncertainty in Medical Expert systems. Fuzzy Sets and Systems. 1983;**11**:197

[10] Buchanan BG, Shortliffe EH. Rule-Based Expert System: The MYCIN Experiments of the Stanford Heuristic Programming Project. Addition-Wesley: Readings, MA; 1984

[11] Klir GJ. Generalized information theory: aims, results, and open problems. Reliability Engineering and System Safety. 2004;**85**(1–3):21-38

[12] Mamdani EH, Assilian S. An experiment in linguistic synthesis with a fuzzy logic controller. International Journal of Human-Computer Studies. 1999;**62**(2):143-147

[13] Sugeno M, Kong GT. Fuzzy modeling and control of multilayer incinerator. Journal of Fuzzy Sets and Systems. 1986;**18**(3):329-345

[14] Rescher N. Many-Valued Logic. New York: McGraw-Hill; 1969

[15] Shafer G. A Mathematical Theory of Evidence. Priceton, NJ: University Press; 1976

[16] Shafer G, Pearl J. Readings in Uncertainy Reasoning. Los Altos, CA: Morgan-Kaufmann; 1990

[17] Reddy PVS. Fuzzy conditional inference for mediiagnosis. In: Second International Conference on Fuzzy Theory and Technology, Proceedings, Abstract and Summaries of FT&T1993; University of North-Carolina/Duke University, USA; October 13–16, 1993. pp. 193-195

[18] Reddy PVS, Babu MS. Some methods of reasoning for conditional propositions. Fuzzy Sets and Systems. 1992;**52**(1):229-250

[19] Reddy PVS. Generalized fuzzy sets in various situations for incomplete knowledge. In: First International Conference on Fuzzy Theory and Technology, Proceedings, Abstract and Summaries of FT&T1992, University of

North-Carolina/Duke University, USA; October 14–16, 1992; pp. 181-183

[20] Reddy PVS. Fuzzy certainty factor for incomplete information. In: 2016 International Conference on Fuzzy Theory and Its Applications (iFUZZY IEEE XPlore.2016); November 9–11, 2016; Taichung, Taiwania

[21] Reddy PVS. Fuzzy conditional inference and application to wireless sensor network fuzzy control systems. In: 2015 IEEE 12th International Conference on Networking, Sensing and Control (ICNSC); IEEE Xplore; 2015. pp. 1-6

[22] Zadeh LA. Calculus of Fuzzy restrictions. In: Zadeh LA, Fu KS, Shimura M, editors. Fuzzy Sets and Their Applications to Cognitive and Decision Processes. New York: Academic; 1975. pp. 1-39

[23] Takac Z. Inclusion and subsethood measure for interval-valued fuzzy sets a nd for continuous type-2 fuzzy sets. Fuzzy Sets and Systems. 2013;**224**(1): 106-120

[24] Yager RR. Uncertainty representation using fuzzy measures. IEEE Transactions on Systems, Man, and Cybernetics, Part B. 2002;**32**: 213-220

[25] Yen J, Langari R. Fuzzy Logic: Intelligence, Control and Information. Prentice Hall of India; 1980

[26] Zadeh LA. A fuzzy-algorithmic sets. Control. 1965;**1**:338-353

Section 5

Expert Systems

Chapter 5

Fuzzy Logic and Fuzzy Expert System-Based Material Synthesis Methods

Mustafa B. Babanli

Abstract

Analyzing a wide diversity of approaches to material selection and synthesis, one can observe a tendency to shift research efforts from physical experiments to systematic analysis based on mathematical models and computational schemes. The latter, in turn, evolves from traditional analytical methods and computational schemes to modern approaches that are based on collaboration of fuzzy logic and soft computing, machine learning, big data and other new methods. In this study, emphasis is put on modeling of fuzzy relationship between performance of new materials and affecting factors. This chapter includes applications of fuzzy model-based synthesis of different alloys. Fuzzy If-then rules based TiNiPt alloy synthesis problem, fuzzy expert system based synthesis of material for pressure vessel and other problems are considered.

Keywords: fuzzy logic, material synthesis, big data, fuzzy clustering, expert system

1. Introduction

Development of new materials is one of important tasks of theoretical and practical interest. Traditionally, this task is implemented mainly on the basis of intensive (and sometimes "ad hoc") experiments which are time- and resource consuming or even not practically implementable. Nowadays, it is well understood that more systematic and effective approaches are needed which are based on computer-guided synthesis of materials. Such approaches rely on data-driven mathematical models and knowledge base obtained from big data previously collected during intensive experiments. Existing computational approaches include methods based on phase diagrams, simulation modeling, theory of associated solutions, methods of microstructure modeling, random fields, etc. In [1], authors analyze the way data-driven techniques are used in deciphering processing-structure-property-performance relationships in materials, with examples of forward (property prediction) and inverse (materials discovery) models. Such analysis can noticeably improve cost-effective materials discovery as the aim of Materials Genome Initiative (MGI). It is shown that adding data sciences to the paradigms of materials science is important to deal with big data.

Agrawal et al. [2] used the Japan National Institute for Materials Science (NIMS) MatNavi database [3] to develop models for prediction of fatigue strength of steel. Prediction accuracy is important for a number of applications due to the significant complexity of fatigue testing and serious consequences of its failures. Actually, fatigue usually leads to more than 90% of all mechanical failures of structural components [4].

In [5], the authors processed the materials properties database for selecting and designing high-temperature alloys for solid oxide fuel cell (or SOFC) applications. Also, this work considers the selection of alloy compositions and properties, which are relevant to the SOFC application. The alloys of interest included such high-temperature alloys as Co, Ni, and Fe base superalloys, as well as stainless steels and Cr base alloys.

The fusion of clustering and regression methods with optimization approaches provides a new opportunity for materials discovery and design. In [6], they discuss the challenges and opportunities associated with materials research. The work [7] for the first time represents machine learning-based determination of viable new compound from true chemical white space, whereas no characterization was provided by promising chemistries. The authors consider an effective prediction model for materials properties that may be easily accessible and useful for researchers.

Existing works based on classical computational schemes used for material synthesis and selection provided good results. However, one important issue related to big data-based computerized material synthesis is that experimental data include measurement errors, partially reliable information, imprecise evaluations, etc. This mandates the use of fuzzy logic approaches for material synthesis. Let us consider some existing works in this regard.

Papers [8–10] show the necessity to account for nonlinearity and uncertainty factors that characterize modeling of material design problems. This requires searching for new ways in formalization of systematic approaches to material design. These papers are devoted to these factors.

Authors in [11] used a new combining tool with which it is possible to model and optimize new alloys that simultaneously satisfy up to 11 physical criteria. To develop a new polycrystalline nickel-base superalloy with the optimal combination of cost, density, gamma-primary phase and sol content, phase stability, durability, yield point, tensile strength, stress rupture, oxidation resistance, and elongation.

In [12], they have developed a rule-based fuzzy logic model for predicting shear strength of Ni-Ti alloy specimens which were produced using powder metallurgy method.

In [13], they applied the fuzzy set theory to knowledge mining from big data on material characteristics. The authors propose fuzzy clustering-generated If-Then rules as a basis for computer synthesis of new materials. These fuzzy If-Then rules describe relationship between material composition and material properties. Validity of the proposed approach is verified on an example of prediction properties of Ti-Ni alloy, and computer experiments of the proposed fuzzy model show its better performance than the physical experiment-based analysis.

In [14], ANFIS model is used to describe the high-temperature deformation behavior of Ni-based superalloy. The inputs of the ANFIS model are deformation temperature, strain rate, and true strain, and the output is true stress. The optimal numbers and types of membership function for the input variables are found. The results show that the constructed ANFIS model is effective in predicting the considered behavior of the Ni-based superalloy.

In this chapter, we propose fuzzy If-Then rule-based model to predict properties of new materials. The model is constructed on the basis of fuzzy clustering of big data on dependence between material composition and related properties. The motivation to use fuzzy model is inspired by the necessity to construct an intuitively well-interpretable development strategy from imperfect and complex data. The proposed approach is applied to synthesis of Ti-Ni-X alloys with required properties and synthesis of material for pressure vessel. Computer experiments of the proposed fuzzy models show better performance than the physical experiment-based analysis.

2. Statement of material synthesis problem and solution methods

The motivation to use fuzzy model is inspired by the necessity to construct an intuitively well-interpretable development strategy from imperfect and complex data. Analyzing a wide diversity of approaches to material selection and synthesis, one can observe a tendency to shift research efforts from physical experiments to systematic analysis based on mathematical models and computational schemes. The latter, in turn, evolutes from traditional analytical methods and computational schemes to modern approaches that are based on collaboration of fuzzy logic and soft computing, machine learning, big data, and other new methods. Uncertainty of materials properties requires to use fuzzy logic methods to more adequately model and predict possible material behavior. This will help to deal with imprecision of experimental data; partial reliability of experimental data, prediction results, and expert opinions; uncertainty of materials properties stemming from complex relationship between material components; and a necessity to analyze, summarize, and reason with a large amount of information of various types (numeric data, linguistic information, graphical information, geometric information, etc.).

Fuzzy logic methods have a good capability to effectively capture and process imprecise experimental data, that is, interpret, classify, learn, and compute with them. Fuzzy logic may help to improve abilities of big data principles to deal with a huge amount and variety of information. In this realm, fuzzy clustering and fuzzy logic-based knowledge bases and information search algorithms provide a bridge between complexity, imperfectness, qualitative nature of information, and research techniques. Particularly, this may help to get intuitive general interpretation of materials science results obtained by various techniques, and ways to get practical results would be then more evident.

Assume that big data on smart materials sourced from experiments is available. These big data describe relationship between alloy composition and its characteristics (**Table 1**) [13, 15, 16].

The problem is to extract knowledge-based model from the considered data and to find an alloy composition which provides a predefined alloy characteristics. We will consider fuzzy knowledge-based synthesis model [17–20]. The problem is solved as follows [21].

First, fuzzy clustering of the big data is applied to determine fuzzy clusters $C_1, C_2, \ldots, C_K$.

Second, fuzzy IF-THEN rule-based model is constructed from $C_1, C_2, \ldots, C_K$:

$$IF\, y_1\, is\, A_{k1}\, and, \ldots, and\, y_n\, is\, A_{kn}\, THEN\, z_1\, is\, B_{k1}\, and, \ldots,$$

Experiment	Alloy composition (in %)			Conditions			Alloy characteristics		
#	Metal 1, y_1	...	Metal n, y_n	Cond.1	...	Cond. l	Char. 1, z_1	...	Char. m, z_m
1	y_{11}	...	y_{1n}	T_{11}	...	T_{1l}	z_{11}	...	z_{1m}
⋮					⋮				
s	y_{s1}	...	y_{sn}	T_{s1}	...	T_{sl}	z_{s1}	...	z_{sm}

Table 1.
Big data of relationship between alloy composition and its characteristics.

$$and\, z_m\, is\, B_{km}, k = 1, \ldots, K \tag{1}$$

Third, fuzzy inference is implemented on the basis of the fuzzy IF-THEN rules to compute optimal values $B'_1, \ldots, B'_m$ of alloy characteristics $z_1, \ldots, z_m$. The fuzzy inference is mainly based on composition of a fuzzy input information on material constituents (and other conditions) and fuzzy relation which describes fuzzy IF-THEN rules. A different approach to fuzzy reasoning also exists and is applied in case of scarce rule base. This is based on fuzzy inference by using similarity of fuzzy input information and antecedents of existing fuzzy rules; a resulting output is then computed as linear interpolation of fuzzy rule consequents.

By using fuzzy inference, optimal values $B'_1, \ldots, B'_m$ are found as those closed to the ideal vector of characteristics $B^* = (B^*_1, \ldots, B^*_m)$. For material synthesis, also fuzzy expert system approach is used. In this case, fuzzy expert system ESPLAN implements IF-THEN rule base obtained from fuzzy clustering of data.

The use of fuzzy rules and fuzzy inference provides us important tools for transition from intensive experiments which deal with a physical model to a fuzzy logic-based mathematical model. Further experiments are conducted not by using physical model but by using fuzzy logic-based mathematical model.

3. Material synthesis of Ti-Ni-X alloys by using ideal vector of characteristics

3.1 Synthesis of Ti-Ni-Pd alloys with given characteristics

A problem of computational synthesis of Ti-Ni-Pd alloy with predefined characteristics is considered. A big data fragment describing dependence alloy composition and the corresponding characteristics is shown in **Table 2**.

A problem of computational synthesis is related to determination of alloy composition with corresponding values of the characteristics close to the target values:

$$z_1 = (302.3), z_2 = (323.3), z_3 = (347.1), z_4 = (331.3) \tag{2}$$

Thus, $B^* = (B^*_1, B^*_2, B^*_3) = ((302.3), (323.3), (347.1), (331.3))$ can be considered as an ideal solution.

In order to describe relationship between alloy composition and the characteristics values, the fuzzy IF-THEN rules were obtained by using FCM clustering of the considered big data:

IF Ni is L and Pd is A2
THEN M_f is A and M_s is A and A_s is aand A_f is A
IF Ni is Aand Pd is A1
THEN M_f is L and M_s is L2 and A_f is L2 and A_s is L
IF N_i is H2 and P_d is L1
THEN M_f is VL and M_s is VL and A_s is Land A_f isVL,
IF Ni is H1 and Pd is L2
THEN M_f is L and M_s is L and A_f is L and A_s is L
IF Ni is VH and Pd is VH
THEN M_f is H and M_s is H and A_f is VH and A_s is VH
The codebooks for inputs are shown in **Tables 3** and **4**.
The linguistic approximation of the inputs is shown in **Tables 5** and **6**.
The codebooks for the outputs are shown in **Tables 7–10**.

Composition			Transformation temperatures			
x_1 (Ni, %)	x_2 (Ti, %)	x_3 (Pd, %)	y_1 (martensitic finish temperature, K)	y_2 (martensitic start temperature, K)	y_3 (austenitic finish temperature, K)	y_4 (austenitic start temperature, K)
41	50	9	322.3	329.4	341.3	331.2
39	50	11	318.2	335.7	347.6	334.7
29	50	21	406.4	424.5	440.3	426.6
20	50	30	515.3	533.8	546.8	534.9

Table 2.
A big data fragment on Ti-Ni-Pd alloy composition [22].

No.	Linguistic value	TFN
1.	Very low (VL)	(3, 3, 13.5) (1)
2.	Low (L)	(3, 13.5, 24) (2)
3.	Average (A)	(13.5, 24, 34.5) (3)
4.	High (H)	(24, 34.5, 45) (4)
5.	Very high (VH)	(34.5, 45, 45) (5)

Table 3.
Codebook for input 1 (Ni).

No.	Linguistic value	TFN
1.	Very low (VL)	(3, 3, 13.75) (1)
2.	Low (L)	(3, 13.75, 24.5) (2)
3.	Average (A)	(13.75, 24.5, 35.25) (3)
4.	High (H)	(24.5, 35.25, 46) (4)
5.	Very high (VH)	(35.25, 46, 46) (5)

Table 4.
Codebook for input 2 (Pd).

No.	Linguistic value	TFN
1.	Very low (VL)	(0, 3.977, 19.2)
2.	Low (L)	(6.709, 18.6, 30.48)
3.	Average (A)	(14.53, 24.7, 34.86)
4.	High 1 (H1)	(21.16, 39.33, 57.51)
5.	High 2 (H2)	(20.88, 30.73, 40.59)

Table 5.
Linguistic terms for input 1 (Ni).

No.	Linguistic value	TFN
1.	Average 1 (A1)	(21.28, 30.03, 38.78)
2.	Average 2 (A2)	(15.9, 24.9, 33.9)
3.	Low 1 (L1)	(0, 10.58, 28.06)
4.	Low 2 (L2)	(9.962, 19.04, 28.13)
5.	Very high (VH)	(28.8, 43.21, 57.62)

Table 6.
Linguistic terms for input 2 (Pd).

No.	Linguistic value	TFN
1.	Average (A)	(394.8, 502.1, 609.5)
2.	Low 1 (L1)	(359.2, 451.3, 543.5)
3.	Very low (VL)	(199.3, 322.3, 445.2)
4.	Low 2 (L2)	(294.4, 386.8, 479.2)
5.	Very high (VH)	(475.4, 674.5, 873.5)

Table 7.
Linguistic terms for output 1 (Mf).

No.	Linguistic value	TFN
1.	Average (A)	(417.4, 523.8, 630.2)
2.	Low 1 (L1)	(369.6, 463.1, 556.6)
3.	Very low (VL)	(221.4, 338.8, 456.2)
4.	Low 2 (L2)	(306.8, 400.4, 494)
5.	Very high (VH)	(532.2, 717.8, 903.5)

Table 8.
Linguistic terms for output 2 (Ms).

No.	Linguistic value	TFN
1.	Average (A)	(414.3, 527.4, 640.5)
2.	Low 1 (L1)	(374.6, 466.5, 558.4)
3.	Very low (VL)	(246.3, 354.8, 463.3)
4.	Low 2 (L2)	(319.1, 409, 498.9)
5.	Very high (VH)	(536.5, 730.6, 924.7)

Table 9.
Linguistic terms for output 3 (As).

The constructed fuzzy model will be used to determine an input vector $A' = \left(A'_1, \dots, A'_n\right)$ that induces the corresponding output vector $B' = \left(B'_1, \dots, B'_m\right)$ maximally close to the ideal solution $B^* = \left(B^*_1, B^*_2, B^*_3\right)$.

We have found that the fuzzy optimal output vector B' induced by the fuzzy input vector $A' = \left(A'_1, A'_2, A'_3\right) = (19.5, 50.5, 30)$ is $B' = \left(B'_1, B'_2, B'_3, B'_4\right) = ((347.78), (364.86), (382.17), (375.22))$. It is the closest vector to the considered ideal fuzzy vector $B^* = ((302), (323), (347), (313))$. The distance is $D\ (B,B^*) = 94$. Thus,

#	Linguistic value	TFN
1.	Average (A)	(420.5, 537.7, 654.9)
2.	Low 1 (L1)	(360.5, 471.1, 581.6)
3.	Very low (VL)	(214.5, 344, 473.6)
4.	Low 2 (L2)	(301.9, 406.6, 511.3)
5.	Very high (VH)	(599.2, 771, 982.8)

Table 10.
Linguistic terms for output 4 (Af).

the computational synthesis based on the fuzzy model uncovers the following alloy composition: Ni is about 19%, Ti is about 51%, and Pd is about 30% with the characteristics $M_f = 347.78$, about $M_s = 364.86$, about $A_f = 382.17$, and $A_s = 375.22$.

3.2 Synthesis of Ti-Ni-Pt alloys with given characteristics

A problem of computational synthesis of Ti-Ni-Pt alloy with predefined characteristics is considered. A big data fragment describing dependence alloy composition and the corresponding characteristics is shown in **Table 11**.

The following fuzzy IF-THEN rules were obtained by using FCM clustering of the considered big data:

If x_1 is VL and x_3 is VH THEN y_1 is VH and y_2 is VH.
If x_1 is H2 and x_3 is L1 THEN y_1 is VL and y_2 is VL.
If x_1 is A and x_3 is L3 THEN y_1 is L2 and y_2 is L2.
If x_1 is L and x_3 is H THEN y_1 is H and y_2 is H.
If x_1 is H1 and x_3 is L2 THEN y_1 is L1 and y_2 is L.

Composition			Transformation temperatures	
x_1 (Ni, %)	x_2 (Ti, %)	x_3 (Pt, %)	y_1 (martensitic start temperature, K)	y_2 (austenitic start temperature, K)
30	50	20	539	544
20	50	30	833	867
15	50	35	953	1023
		…		
10	50	40	1173	1123

Table 11.
Transformation temperatures of Ti-Ni-Pt alloy [23].

No.	Linguistic value	TFN
1.	Very low (VL)	(5, 5, 13.75)
2.	Low (L)	(5, 13.75, 22.5) (2)
3.	Average (A)	(13.75, 22.5, 31.25) (3)
4.	High (H)	(22.5, 31.25, 40) (4)
5.	Very high (VH)	(31.25, 40, 40) (5)

Table 12.
Codebook for input 1 (x_1).

No.	Linguistic value	TFN
1.	Very low (VL)	(10, 10, 18.75) (1)
2.	Low (L)	(10, 18.75, 27.5) (2)
3.	Average (A)	(18.75, 27.5, 36.25) (3)
4.	High (H)	(27.5, 36.25, 45) (4)
5.	Very high (VH)	(36.25, 45, 45) (5)

Table 13.
Codebook for input 2 (x_3).

No.	Linguistic value	TFN
1.	Very low (VL)	(0, 7.535, 22.65)
2.	High 1 (H1)	(25.98, 35.17, 44.35)
3.	Average (A)	(19.27, 26.33, 33.39)
4.	Low (L)	(8.109, 17.64, 27.17)
5.	High 2 (H2)	(21.67, 30.06, 38.48)

Table 14.
Linguistic terms for input 1 (Ni).

No.	Linguistic value	TFN
1.	Very high (VH)	(27.18, 42.47, 57.75)
2.	Low 1 (L1)	(5.859, 14.84, 23.82)
3.	Low 2 (L2)	(14.14, 21.78, 29.42)
4.	High (H)	(22.63, 32.36, 42.08)
5.	Low 3 (L3)	(13.71, 19.75, 26.18)

Table 15.
Linguistic terms for input 2 (Pt).

No.	Linguistic value	TFN
1.	Very low (VL)	(363, 363, 565.5)
2.	Low (L)	(363, 565.5, 768)
3.	Average (A)	(565.5, 768, 970.5)
4.	High (H)	(768, 970.5, 1173)
5.	Very high (VH)	(970.5, 1173, 1173)

Table 16.
Codebook for output 1 (y_1).

The codebooks for inputs are shown in **Tables 12** and **13**.

The linguistic approximation of the inputs is shown in **Tables 14** and **15**.

The codebooks for the used outputs are shown in **Tables 16** and **17**.

We have found that the fuzzy optimal output vector B' induced by the fuzzy input vector $A' = \left(A'_1, A'_2, A'_3\right) = (40, 50, 10)$ is $B' = ((479.68), (488))$. It is the closest vector to the considered ideal fuzzy vector $B^* = ((363), (373))$. The distance between them is $D(B', , B^*) = 164$. The fuzzy model-based results show that the

No.	Linguistic value	TFN
1.	Very low (VL)	(373, 373, 585.5)
2.	Low (L)	(373, 585.5, 798)
3.	Average (A)	(585.5, 798, 1010.5)
4.	High (H)	(798, 1010.5, 1223)
5.	Very high (VH)	(1010.5, 1223, 1223)

Table 17.
Codebook for output 2 (y_2).

desired alloy composition is as follows: Ti is about 50%, Ni is about 37%, Pt is about 13%, and the obtained characteristics are about $M_s = 479{,}6828$ and about As = 488,1005.

4. Material synthesis by fuzzy expert system

A series of works exist on material synthesis by using fuzzy models [12, 24, 25]. In this study, to solve material synthesis problem for pressure vessel, two methods are used: possibility measure-based inference method (by ESPLAN shell, Aliev inference) and Mamdani inference method (by MATLAB environment, Fuzzy Toolbox) [26].

4.1 Statement of the problem

Defining the performance index for pressure vessel in material synthesis is a very important problem. The basic problem is to evaluate the performance index by using weighted performance indices.

For determining the performance index, we use data of alloys. There are many types of alloys.

The weighted performance index denoted *Out* is a compound index built from four characteristics each of which is extracted from the data set. The four characteristics are *in1*-scaled PREN, *in2*-scaled yield strength, *in3*-scaled weldability, and *in4*-scaled impact strength.

Using the abovementioned parameters, the performance index model can be expressed as.

IF x_1 is A_{11} and ..... x_n is A_{1n} THEN y is B_1.

IF x_1 is A_{21} and ..... x_n is A_{2n} THEN y is B_2.

... ...

IF x_1 is A_{m1} and ..... x_n is A_{mn} THEN y is B_m.

where $x_j = {}^j 1...n$ are the linguistic input variables, y is the output variable, and A_{ij} and B_i are the fuzzy sets, n = 4, m = 7.

Fragment of data set is given in **Table 18**.

4.2 Modeling of material data by fuzzy C-means clustering

To create this model, we use clustering approach, mainly fuzzy C-means. Data set contains 35 records extracted from big data. For modeling we use two-thirds of the given data and testing one-third. Inputs: x1, scaled PREN; x2, scaled yield; x3, scaled weldability; x4, scaled impact strength. Output: y, performance index. For simulation FCM-based clustering initial data are:

Scaled PREN	Scaled yield strength	Scaled weldability	Scaled impact strength	Performance index
26.60	3.60	18.40	5.00	53.50
29.70	4.40	23.00	8.60	65.60
19.80	3.60	23.00	5.00	51.30
22.30	3.20	23.00	8.60	57.10
26.00	3.60	18.40	6.80	54.70
22.30	5.40	13.80	11.30	52.70
...	...	...	...	...
47.00	4.60	18.40	13.50	83.50
29.70	4.40	18.40	15.80	68.30
20.40	12.00	18.40	5.00	55.80
21.00	9.80	23.00	4.50	58.30
23.50	4.60	23.00	13.50	64.60
11.80	2.50	18.40	9.00	41.60
15.50	2.50	18.40	8.80	45.10
22.90	5.80	13.80	7.10	49.50
26.60	6.20	4.60	3.20	40.50
...	...	...	...	...
18.60	2.90	18.40	8.80	48.60
32.20	6.20	18.40	6.00	62.70
42.70	4.30	23.00	15.20	85.10
21.00	2.50	18.40	8.80	50.70
21.60	9.50	18.40	4.50	54.00

Table 18.
Fragment of data set (extracted from big data).

Cluster numbers = 7.
Max iteration =1000.
Exponent = 2.
Min. improvement = 0.000001.

Obtained centers of the clusters are given in **Table 19**. Each row describes a cluster center as five-dimensional vector with coordinates x1 (scaled PREN), x2 (scaled yield), x3 (scaled weldability), x4 (scaled impact strength), and y (performance index). Columns describe the values of the coordinates of the cluster centers.

Representation of the extracted fuzzy rules from big data by using fuzzy c-means method fragment is given below and in **Figure 1**.

1. *IF Scaled PREN = about 18 and Scaled yield = about 3 and Scaled weldability = about 14.5 and scaled impact strength = about 10.8, THEN Performance index = about 46.5.*

2. *IF Scaled PREN = about 27 and Scaled yield = about 4.4 and Scaled weldability = about 21 and scaled impact strength = about 12 THEN Performance index = about 65.*

3. *IF Scaled PREN = about 26 and Scaled yield = about 5 and Scaled weldability = about 4.8 and scaled impact strength = about 3 THEN Performance index = about 38.5.*

4. *IF Scaled PREN = about 21 and Scaled yield = about 9 and Scaled weldability = about 21.2 and scaled impact strength = about 5 THEN Performance index = about 55.*

5. *IF Scaled PREN = about 25 and Scaled yield = about 3.6 and Scaled weldability = about 19 and scaled impact strength = about 6 THEN Performance index = about 53.5.*

6. *IF Scaled PREN = about 47 and Scaled yield = about 4.5 and Scaled weldability = about 18 and scaled impact strength = about 13 THEN Performance index = about 83.*

	x1	x2	x3	x4	y
Center 1	18.5215	3.0384	14.7548	10.7647	46.9898
Center 2	27.6395	4.4574	21.3502	12.6412	66.0559
Center 3	26.0329	4.9933	4.8428	3.2598	39.0312
Center 4	20.8528	9.7068	21.3952	5.0337	56.9826
Center 5	25.0287	3.5955	19.0063	6.3912	53.9372
Center 6	46.9418	4.5996	18.4040	13.4963	83.4416
Center 7	24.1544	6.2895	14.0802	9.7481	54.2839

Table 19.
Centers of the clusters.

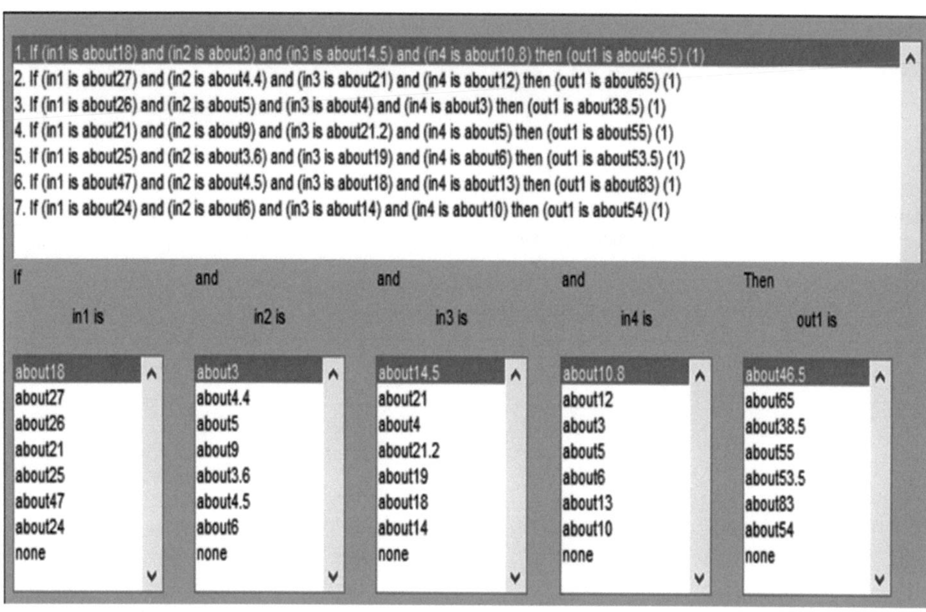

Figure 1.
Extracted fuzzy rules (by using fuzzy C-means method).

7. *IF Scaled PREN = about 24 and Scaled yield = about 6 and Scaled weldability = about 14 and scaled impact strength = about 10 THEN Performance index = about 54.*

Graphical representation of the linguistic terms of inputs and outputs of the rules as trapezoidal fuzzy numbers is given in **Figures 2–6**.

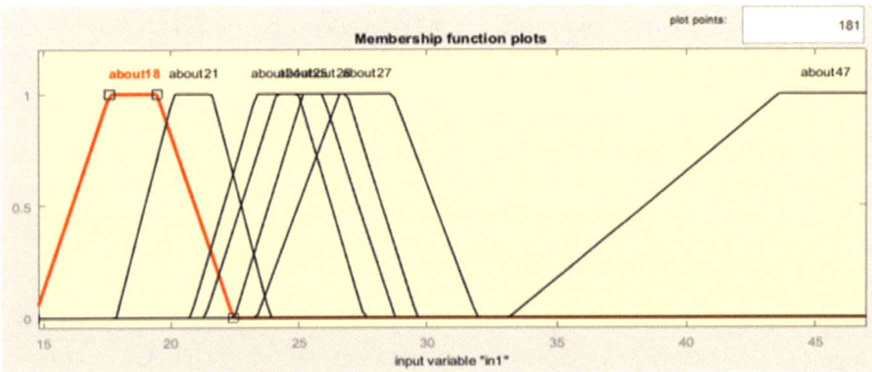

Figure 2.
Linguistic terms of input 1 (scaled PREN).

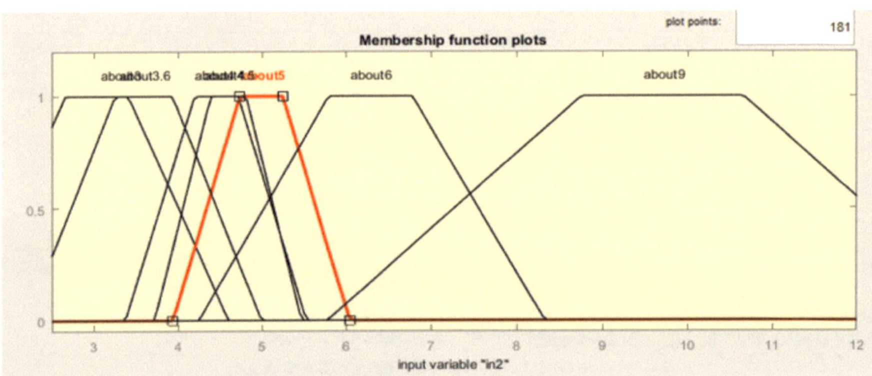

Figure 3.
Linguistic terms of input 2 or scaled yield strength.

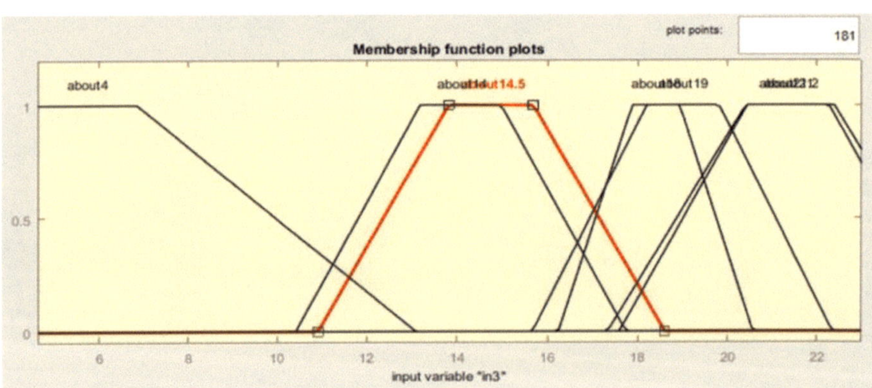

Figure 4.
Linguistic terms of input 3 or scaled weldability.

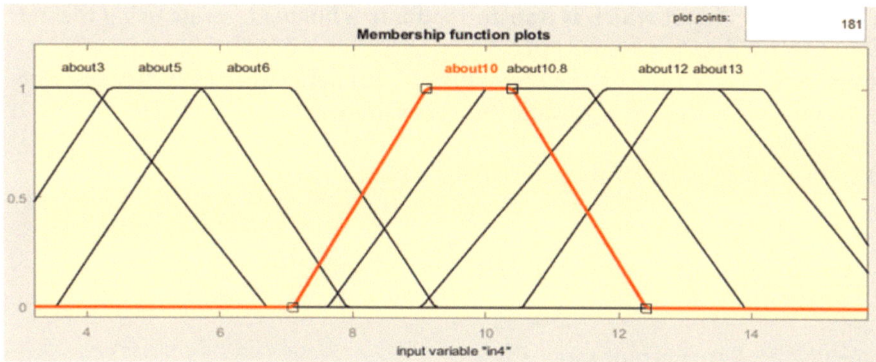

Figure 5.
Linguistic terms of input 4 or scaled impact strength.

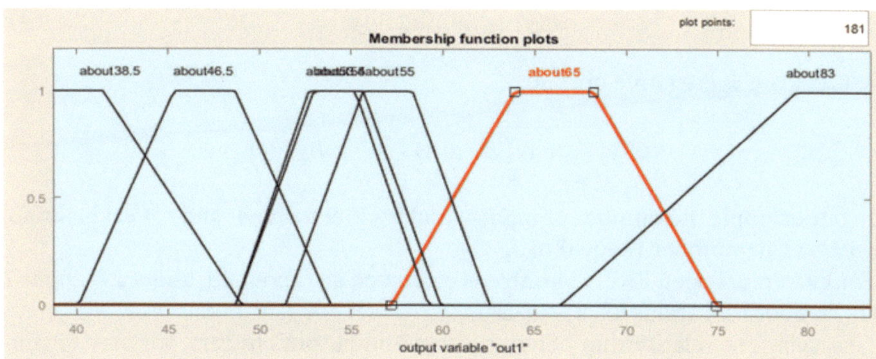

Figure 6.
Linguistic terms of outputs or performance index.

4.3 Solution of the problem

For solving the problem described in Section 4.1, we will use ESPLAN shell.

The problem is to determine material with the given level of performance index using the fuzzy model obtained in Section 4.2.

In this context we define basic objects and linguistic terms according to ESPLAN shell. The linguistic terms are described by trapezoidal fuzzy numbers. The rule base given above is put as knowledge base in ESPLAN shell. Then, different tests are performed.

TEST 1.

IF Scaled PREN = about 18 and Scaled yield = about 3 and.

Scaled weldability = about 21 and scaled impact strength = about 12.

THEN Performance index =?

ANSWER:

EXPERT system shell ESPLAN's result is "Performance index is about 46.5" **(alloy Monel-400)**.

The fuzzy rules were derived from alloy big data by using FCM method, and fuzzy inference within these rules is implemented in expert system shell ESPLAN. The obtained results confirm efficiency of the proposed approach.

Solution by using Mamdani inference method. General form of the abovementioned rules are as form (4.13). Mamdani fuzzy inference is most commonly used approximate reasoning methodology for fuzzy modeling. The method works with crisp input which is transformed into a linguistic value using the antecedent membership functions. After the aggregation process of consequents induced by antecedents, obtained final fuzzy set is defuzzified. We can describe fuzzy inference process in algorithmic view as follows:

1. Firing level for each rule is defined as follows:

$$\alpha_i = \min_{j=1}^{n} \left[\max_{x_j} \left(A_j^{'}(x_j) \wedge A_{ij}(x_j) \right) \right] \tag{3}$$

where $A_j^{'}(x_j)$ are current input values.

2. Outputs for each rule are calculated:

$$B_i^{'}(y) = \min(\alpha_i, B_i(y)) \tag{4}$$

3. Calculate aggregative output:

$$B^{'}(y) = \max\left(B_1^{'}(y), B_2^{'}(y), \ldots., B_m^{'}(y)\right) \tag{5}$$

In our example the number of input variables is equal to 4, and for each variable, linguistic value number is equal to 7.

For example, scaled PREN variable is evaluated as (about 18, about 27, about 26, about 21, about 25, about 47, about 24).

Observing the relationship between input and output clusters, we may formulate the following linguistic descriptions—productions rules, for example:

1. *IF In1 about 18 and In2 = about 3 and In3 = about 14.5 and In4 = about 10.8 THEN Out = about 46.5.*

2. *IF In1 = about 27 and In2 = about 4.4 and In3 = about 21 and In4 = about 12 THEN Out = about 65.*

3. *IF In1 = about 26 and In2 = about 5 and In3 = about 4.8 and In4 = about 3 THEN Out = about 38.5.*

4. *IF In1 = about 21 and In2 = about 9 and In3 = about 21.2 and In4 = about 5 THEN Out = about 55.*

5. *IF In1 = about 25 and In2 = about 3.6 and In3 = about 19 and In4 = about 6 THEN Out = about 53.5.*

6. *IF In1 = about 47 and In2 = about 4.5 and In3 = about 18 and In4 = about 13 THEN Out = about 83.*

7. *IF In1 = about 24 and In2 = about 6 and In3 = about 14 and In4 = about 10 THEN Out = about 54.*

The obtained rules are put into Fuzzy Toolbox to perform tests by using the following data (**Table 20**):

Below, we provide some test results.

Test results. The following input data are given:

Scaled PREN	Scaled yield strength	Scaled weldability	Scaled impact strength	Performance index
18.60	2.90	18.40	8.80	48.60
21	2.5	18.4	8.8	50.7

Table 20.
Testing data (fragment).

In1 = 18.60, *In2* = 2.90, *In3* = 18.40 and *In4* = 8.

For these data, the following defuzzified output describing the alloy performance index is computed by using Mamdani fuzzy inference:

Out = 48.60.

This value fits the performance index of alloy 317 L (from the given big data).

Consider other values of the inputs:

In1 = 18.60, *In2* = 2.90, *In3* = 18.40 and *In4* = 8.80.

For these values, the defuzzified output is *Out* = *50.3*. This result fits the performance index of alloy 317LM.

Consider also the following input values:

In1 = *26*, *In2* = *3.6*, *In3* = 18.4, and *In4* = 6.8.

The computed output (performance index) is 54.7. The performance index computed for the third case and the performance index from big data set are shown in **Table 21**.

Deviation between testing and expert data is 0.18% or 0.0018.

Summarizing the findings in this chapter, we have to conclude that the discovery and design of new materials are driving forces for much of the research that takes place in multiple disciplines, including materials science and engineering, matter physics, materials chemistry, and emerging technologies such as fuzzy logic, soft computing, etc. However, this task is implemented mainly on the basis of time- and resource-consuming experiments. Thus, we consider to shift the approaches to material design investigations from physical experiments to experiments on the basis of fuzzy If-Then rule-based material model. The motivation to use fuzzy model is inspired by the necessity to construct an intuitively well-interpretable material design model based on imperfect and complex data. In this chapter we have considered three material synthesis problems which had shown that instead of carrying out complicated experiments, researchers can use fuzzy model-based computational synthesis approach utilizing digital twins of physical models. Applications of this approach have shown that fuzzy model-based experiments can give better results than physical experiments in terms of desirable characteristics of synthesized materials. The approaches suggested in this chapter are universal and may be applied not only in materials science but also in chemical engineering, drug design, and other fields. Complexity of material design problems mandates to combine fuzzy logic and efficient learning methods as artificial neural networks, evolutionary algorithms, and others to more adequately model and predict possible material behavior.

Scaled PREN	Scaled yield strength	Scaled weldability	Scaled impact strength	Computed performance index	Given performance index
26.00	3.60	18.40	6.80	54.70	54.8 (alloy 1925hMo)

Table 21.
Comparison of computed and given data.

5. Conclusion

In this chapter, we have used data-driven approach to construction of fuzzy model which is more effective than expert-driven approach. Consequently, we have used fuzzy C-means clustering to derive fuzzy If-Then rules from material data that describe material composition and related characteristics. In order to determine the required characteristics, computational experiments on the basis of fuzzy inference and fuzzy expert system were conducted. The expert opinions and some few physical experiments have proven validity of the obtained results. The main advantage of the fuzzy logic-based approach is a high interpretability of fuzzy If-Then rules. However, learning the ability of the fuzzy models is scarce. Thus, combination of fuzzy logic with deep learning methods, mainly, reinforcement learning methods, would help to achieve better results on material synthesis.

In future works, fuzzy materials paradigm may improve processing-structure--property-performance relationship in hierarchy of structural materials levels, from the atomic and electronic to the macrostructural levels. Another important application of fuzzy logic is fuzzy phase diagram construction for different alloy models using uncertain enthalpies and other thermodynamic parameters will be investigated, which opens a door to design new materials.

Author details

Mustafa B. Babanli
Azerbaijan State University of Oil and Industry, Baku, Azerbaijan

*Address all correspondence to: mustafababanli@yahoo.com

References

[1] Agrawal A, Choudhary A. Perspective: Materials informatics and big data: Realization of the 'fourth paradigm' of science in materials science. APL Materials. 2016;**4**:1-10

[2] Agrawal A, Deshpande PD, Cecen A, Basavarsu GP, Choudhary AN, Kalidindi SR. Exploration of data science techniques to predict fatigue strength of steel from composition and processing parameters. Integrating Materials and Manufacturing Innovation. 2014;**3**:1-19

[3] National Institute of Materials Science. Available from: http://smds.nims.go.jp/fatigue/index_en.html [Accessed: January 12, 2016]

[4] Dieter GE. Mechanical Metallurgy. New York: McGraw-Hill Book Company; 1986

[5] Yang ZG, Stevenson JW, Paxton DM, Singh P, Weil KS. Materials Properties Database for Selection of High-Temperature Alloys and Concepts of Alloy Design for SOFC Applications. Richland, Washington: Pacific Northwest National Laboratory; 2002. 78 p

[6] Hill J, Mulholland G, Persson K, Seshadri R, Wolverton C, Meredig B. Materials science with large-scale data and informatics. Unlocking new opportunities. MRS Bulletin. 2016; **41**:399-409

[7] Gaultois MW et al. Perspective: Web-based machine learning models for real-time screening of thermoelectric materials properties. APL Materials. 2016;**4**(5):053213-1-053213-11. DOI: 10.1063/1.4952607

[8] Elishakoff I, Ferracuti B. Fuzzy sets based interpretation of the safety factor. Fuzzy Sets and Systems. 2006;**157**: 2495-2512

[9] Lee Y-H, Kopp R. Application of fuzzy control for a hydraulic forging machine. Fuzzy Sets and Systems. 2001; **99**:99-108

[10] Rao HS, Mukherjee A. Artificial neural networks for predicting the macromechanical behaviour of ceramic-matrix composites. Computational Materials Science. 1996;**5**:307-322

[11] Conduit BD, Jones NG, Stone HJ, Conduit GJ. Design of a nickel-base superalloy using a neural network. Materials & Design. 2017;**131**:358-365

[12] Tajdari M, Mehraban AG, Khoogar AR. Shear strength prediction of Ni–Ti alloys manufactured by powder metallurgy using fuzzy rule-based model. Materials and Design. 2010;**31**:1180-1185

[13] Babanli MB. Synthesis of new materials by using fuzzy and big data concepts. Procedia Computer Science. 2017;**120**:104-111

[14] Chen DD. Dislocation substructures evolution and an adaptive-network based fuzzy inference system model for constitutive behavior of a Ni-based super alloy during hot deformation. Journal of Alloys and Compounds. 2017; **708**:938-946

[15] Babanli MB, Huseynov VM. Z-number-based alloy selection problem. Procedia Computer Science. 2016;**102**: 183-189

[16] Babanli MB, Kolomytsev V, Musienko R, Sezonenko A, Ochin P, Dezellus A, et al. Thermodynamic properties and thermal stability of the multicomponent TiNi based alloy ribbons. Metal Physics and Advanced Technologies. 2001;**23**:111-124

[17] Zadeh LA. Fuzzy sets. Information and Control. 1965;**8**:338-353

[18] Babanli MB. Theory and practice of material development under imperfect information. In: Advances in Intelligent Systems and Computing. Vol. **896**. Springer; 2018. pp. 4-14

[19] Babanli MB, Prima F, Vermaut P, Demchenko LD, Titenko AN, Huseynov SS, et al. Material selection methods: A review. Advances in Intelligent Systems and Computing. 2018;**896**: 929-936

[20] Babanli MB. Fuzzy modeling of phase diagram under imprecise thermodynamic data. In: Proceedings of the Tenth World Conference "Intelligent Systems for Industrial Automation". b-Quadrat Verlag; 2018. pp. 265-266

[21] Babanli MB, Prima F, Vermaut P, Demchenko LD, Titenko AN, Huseynov SS, et al. Review on the new material design methods. Advances in Intelligent Systems and Computing. Cham, Switzerland: 2018;**896**:937-944

[22] Frenzel J, Wieczorek A, Opahle I, Maa B, Drautz R, Eggeler G. On the effect of alloy composition on martensite start temperatures and latent heats in Ni–Ti-based shape memory alloys. Acta Materialia. 2015;**90**:213-231

[23] Vafaeenezhad H, Seyedein SH, Aboutalebi MR, Eivani AR. Application of constitutive description and integrated ANFIS – ICA analysis to predict hot deformation behavior of Sn-5Sb lead-free solder alloy. Journal of Alloys Compounds. 2017;**697**:287-299

[24] Odejobi OA, Umoru LE. Applications of soft computing techniques in materials engineering: A review. African Journal of Mathematics and Computer Science Research. 2009;**2** (7):104-131

[25] Datta S, Chattopadhyay PP. Soft computing techniques in advancement of structural metals. International Materials Reviews. 2013;**58**:475-504

[26] Aliev RA, Aliyev RR. Soft Computing and Its Applications. New Jersey, London, Singapore, Hong Kong: World Scientific; 2001

Section 6

Control of Electrical Systems

Chapter 6

Determination of Optimal Transformation Ratios of Power System Transformers in Conditions of Incomplete Information Regarding the Values of Diagnostic Parameters

Lezhniuk Petro Demianovych,
Rubanenko Oleksandr Evgeniovich
and Rubanenko Olena Oleksandrivna

Abstract

On the base of damage rate analysis of power transformers and methods of electrical energy system (EES) modes control the necessity of using the results of on-line diagnostics of LTC transformers not only for determinations of the expending of further operation or equipment repair but also for calculation of optimal transformation coefficients (with account of the suggested RRCT) for their application in the process of modes control has been proved. Improved method of determination of control action, realized by the LTC transformers by means of comparative analysis of the results calculation of EES modes with quasi resistances of the circuit branches. Such peculiarity of the suggested method of determination of control actions by LTC transformers, as the account of RRCT, in the process of EES mode control provides such advantages as reduction of the damage rate of the equipment, reduction of active power losses in EES. Due to the peculiarities of the method of determination of control actions by LTC transformers, with the account of their technical state, perspectives of the development and introduction in EES modern microprocessor-based systems of automatic control of transformers LTC open.

Keywords: on-line diagnostics, control, normal modes, active power losses, similarity theory, neuro-fuzzy modeling, basic similarity criteria, membership functions, uncertainty

1. Introduction

Characteristic feature of present day situation are the attempts of utility companies to increase energy efficiency in conditions of continuing aging of high voltage equipment.

Practice of large-scale introduction of the intelligent support of decision making of solution processes proves their efficiency. One of the directions of efficiency increase of electric energy transportation is improvement of methods and means of active power losses reduction on conditions of maintaining reliable operation of high voltage equipment, including outdated equipment.

The set of electric energy system (EES) states and the processes of transition from one state into another her is EES mode (further-mode), characterized by the parameters, for instance, electrical voltages an substations loads, currents in transmission lines transformation ratios of the transformers, etc., normal operation mode 07 of EES modes. Control by the system of on-line dispatching control (LDC). Modern LDS technologies, for instance, provided by smart grids concept, are aimed at improvement of its information support. This enables the optimal implement more efficiently energy-saving technologies in electric systems, when out of date high-voltage equipment is using.

Means of similarity theory, in particular the criterion method (KM), can effectively solve and analyze optimization problems [1]. Criteria-based method can be defined as a set of techniques and principles, according to which the analysis, comparison and interpretation of baseline data to provide scientific and practical conclusions. The ultimate goal of studies using the criterial method is to reveal regularities, which under certain conditions can be represented as the law control [2]. The main purpose of KM is to find the variant of the process or the object. Most often, in the further analysis, the parameters are used as reference. According to his idea, KM is close to the geometrical programming [3]. This use of duality of optimization problems is the replacement of the direct problem for the corresponding dual. The main difference is that the basis for geometric programming is inequality between geometric and arithmetic averages, and the background of the KM—matrix properties of dimensions or indicators [4]. This is the meaning of dual variables. In geometric programming are weighting factors, in KM—similarity criteria. That is, the result of solving the tasks of the KM is values of criteria of similarity or, in other words, the optimum ratio of the individual parameters, and not they themselves. This specific feature characteristic only of KM, determines its scope.

2. Analysis of the literature data and problem set up

In [5] the technique of voltage drop decrease in separate parts of distribution electric grids is, suggested, but it does not take into account technical state of regulating devices [6]. In [7] on the example of Indian electric grids the statistics of the increase of power, transmitted in electric grid is considered.

It is stated that in order to improve the grid reliability and efficiency of energy transmission it is necessary to use power transformers, equipped with LTC and automatic or automated control systems for such LTC control. It enables to control power flows in EES by means of LTC so that the parameters of electric grids modes were within the limits of normal values of equipment (transmission lines, switching devices, transformers, etc.) parameter.

It is provided by usage of FACTS technologies. For instance, in [8], the possibilities of using phase-shift transformer for power flows change in electric energy system to reduce power losses in the process of energy transmission in transmission lines of Slovak Republic are considered. In [9] high price of FACTs technologies usage in energy branch for the reduction of electrical power is proved. In [10] three variants of power flows control in electric grid, using three FACTs devices are considered with Thyristor Controlled Series Capacitor (TCSC), Static Synchronous Series Compensator (SSSC) and phase-shift transformer (PST), but in [11] any attention is paid to the state of equipment, used for modes control. In [12] the conclusion is made that prolongation of power transformers operation term for

20–30 years is more profitable than their replacement by new ones, and the quantity of power transformers in the USA, that have been in operation for more than 25 years (certificate resource – 25 years) is approximately 65%.

In [13] attention is paid to the system of continuous monitoring of technical state of power transformers, the given system is used at the transformers of joint-stock company "Magnitogorsk Metallurgical Complex", HYDRAN analyzer, methods of localization and identification of faults, practical necessity of partial discharges control is underlined, but the results of diagnostics during modes control are not paid attention to.

In [14] it is noted that in local electric system in order to provide stable operation and indices of electric energy quality it is necessary to use modern control systems that take into consideration voltages in nodes and frequency and eliminate emergency deviations. At the same time, in [15] modeling of non-stationary critical operation modes of EES in the process of parameters change in wide limits by means of application of non-linear mathematical models attention is paid to. This enables to study the consequences of such modes, promptly take measures, aimed at their prevention or elimination. In pages [16] technical state of the equipment of these systems is not taken into account, this can lead to the damage of the equipment and undersupply of energy to the consumers.

Thus, the problem of development of the methods of diagnostics results account during control of EES modes is not solved.

It is known that operation control (RTOC) in Ukraine is carried by a man. Overloading of this person with a great volume of diagnostic parameters data, especially in conditions of limited time for decision-making, leads to their actions. In the process of modes on-line control, especially post-occident modes. It is expedient to assess the state of equipment by generalized indices, for instance, by residual resources coefficient of the transformers (RRCT). The development of the method of on-line diagnostics of the transformers and the account of RRCT in the process of ESS modes control for minimization of total losses of active power are not considered in literature sources and is the subject of authors study.

3. Objectives and tasks of the research

The objective of the research is the development of the method of diagnostics of the transformers with LTC and account of PRCT values in the process of EES modes control for minimization of active power losses. To realize this objective the following problems are to be solved:

- substantiate the expediency of applying the results of diagnostics of LTC-transformers in the process of optimal control of EES modes;
- develop neuro-fuzzy model of residual resource coefficient of the transformers (RRCT);
- develop the method of RRCT values and power transformers with LTC state account in the process of EES modes control.

4. Materials and methods of transformers diagnostic study

Automation of the process of power flow control may be provided by means of centralized remotely controlled alternative usage of switching devices (LTC) of the transformers. Under such conditions, there appears the possibility of the analysis of control actions of separate LTC on mode parameters of EES by means of the feedback. This approach improves the operation quality of adaptive control automatic systems of the LTC position control. For this purpose, at considerable changes of load schedule it is necessary to perform ranking of the transformers with LTC by the quality of their impact on maintaining parameters of the modes.

Realization of measures, aimed at reduction of power losses is limited by the possibilities of the equipment involved in the provision of mode; namely, by its technical state. It is known, that the damage of high voltage equipment during mode control (for instance, power transformers) leads to losses, which considerably exceed the cost of electric energy, saved as a result of losses decrease. Failure rate of the outdated high voltage equipment (power transformers, shunting reactors, instrument current and voltage transformers, switches, etc.) increases, when such equipment has been in operation for more than 25 years [17]. Taking into consideration the fact that the control of EES modes is accompanied by the operation of switching devices, regulation devices of transformers, emergence of switching surges, ferro-resonances, currents increase in power and instrument transformers, transmission lines, etc., then the control of modes must be realized, taking into consideration their technical state [18] and possible expenses for their replacement or repair.

Thus, it is necessary to know current state of high voltage electric equipment of EES, which is in operation during modes control.

5. Determination of current technical state of power transformers

We will consider the method of determination of power transformers current state and RRCT values in the process of EES modes control on the example of power high voltage transformers, which have on-load-tap changing device.

We suggest to evaluate technical state of power transformer by means of the analysis of the value of its residual resource coefficient. Power transformer residual resource coefficient has the dimensionality in relative units and can change in the process of operation in the range from one (the best technical state) to zero (the worst technical state, when the transformer must be removed out of service for inspection, repair, replacement, etc.).

Then, we will consider the example of residual resource coefficient determination of the transformer ATDCTN 125000–330/110. First we will study the statistics of failure rate of such transformers. **Table 1** contains the example of possible reasons and amount of transformers removal out of service, that is close to data, published in studies [19].

In **Table 1** such symbols are used: Z_k is the resistance of the transformer windings (during measurements in short – circuit mode); t° is the temperature of contact points (for instance, bushing of the bus duct or with winding lead); $P_{i.p.}$ idle mode power, that characterizes the quality of magnetic circuit; R_{in} is the resistance of the insulation for revealing the contamination and aging of solid and liquid insulation (also it is necessary to determine the capacity and dielectric loss tangent, also it is desirable to determine the degree of polymerization); W humidification of the isolation; $k_{resid.res.bush}$ or k_{bush} is the residual resource coefficient of the bushings; $CADG_C$ is residual resource coefficient of the transformer by the results of chromatographic analysis of dissolved gas in the transformer oil of the tank and LTC (ethylene, ethane, methane) of the transformer, that characterizes oil contamination by the gases, dissolved in it and among them acetylene and hydrogen (for revealing of discharges); PCA residual resource coefficient of the transformer by the results of physical–chemical analyses of transformer oil from transformer tank,

Transformer element	Designation	Parameter name	Units	%
Windings	Z_k	Winding deformation	8	1.6
	t^0	Deterioration of contact joints state	10	2
	$P_{i.p}$	Idle power that characterizes of the magnetic quality	15	3
Insulation	$R_{in}R$	Contamination of isolation	65	13.4
	W	Humidification of the isolation	48	10
Bushings	k_{bush}	Defects of bushings	74	15.2
Oil	$CADG_c$	Content of dissolved gases	71	14.6
	PCA	High moisture content and deviations of other parameters of the oil	43	9
	$CADG_d$	Discharges in oil	64	13.2
LTC	$k_{def.LTC}$	LTC defects	45	9.3
Cooling system	I_{motor} or I_{mt}	The current of oil pump drive motor	14	2.9
	t^o_{cool}	Coolers temperature	16	3.3
Tank	k_{tank}	Tank leakage	12	2.5
Total			485	100

Table 1.
Reasons of removing out of service power transformers.

contactor and LTC tap changer; $CADG_d$ is residual resource coefficient of the transformer by the results of chromatographic analysis of the dissolved hydrogen and acetylene in the transformer oil of the tank and LTC of the transformer tap changer in order to reveal the discharge; $k_{def.LTC}$ or k_{LTC} coefficient of the transformer LTC residual resource; I_m is the current of electric motors oil pumps and fans of cooling system; $t^o{}_{cool}$ is coolers temperature; k_{tank} is the residual resource coefficient of the transformer tank, determined by the availability (takes the value "0") or absent of oil leakage (takes the value 1).

From **Table 1** shows that the transformers are often displayed in repairs due to moisture and oil contamination, insulation and high-voltage inputs defects.

The task of creating a mathematical model complicated with incomplete initial data as part of the parameters known at the time of payment, such as the reasons for the need for additional studies. To establish reciprocal links diagnostic parameters very constructive simulation technology is unclear. This simulation allows to obtain more reliable results compared to the results of existing diagnostic systems.

In **Table 1** under the term the controlled diagnostic parameter we mean the parameter deviation of which from the norm helped to remove the transformer out of service or was taken into account in the process of its removal out of service. In **Table 1** the following diagnostic parameters are given: parameters, that characterize the state of the windings, insulation, bushings, oil, LTC, cooling systems, tank.

Having analyzed the data of **Table 1** the scheme was created that shows whether dependent or independent is the impact of diagnostic parameters on the coefficient of total residual resource of the transformer (**Figure 1**).

Figure 1 does not show mutual impact of one controlled diagnostic parameter on the other one; it is shown either in dependent or independent manner how these parameters influence the coefficient of total residual resource of power transformer (PT).

In **Figure 1** over the parameter the percentage amount of revealed faulty transformers by the given parameter is shown, that is given in percent from the total amount of faulty transformers.

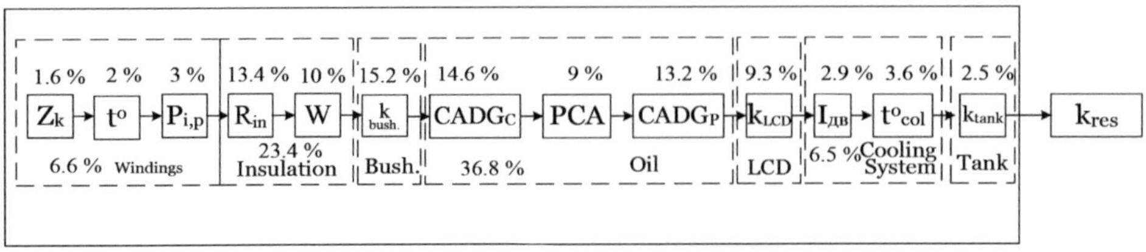

Figure 1.
Structural diagram of the model of total residual resource coefficient of the transformer.

The blocks with parameters whose deviations from the norm substantiate the necessity of the output of the transformers for repair, are shown sequentially and are shown k_{res} – residual resource coefficient of the transformer (RRCT). In parallel, blocks with parameters are also depicted. A large change in these parameters proves the necessity of outputting a power transformer (PT) for repair. PT is repaired in case of deviation from the norms of these parameters. This is due to the requirements for the reliability of the transformers. In each of the given blocks parallel can be allocated but it are not shown to simplify the calculation (for instance, currents of electric motors of oil pumps and fans).

In order to obtain the generalized parameter of the residual resource of the transformer, it is proposed from the known values of diagnostic parameters to pass to the corresponding values of residual resources coefficients (in relative units) by each diagnostic parameter. This will allow you to take into account the value of all diagnostic parameters and the impact of each of them.

These coefficients are defined in relative units by (1) and that is why they characterize total output of the transformers from the moment of their technical state control to transition to boundary state that is residual technical resource (12). Residual resource coefficient k_{i_1} by i_1th diagnostic parameter:

$$k_{i_1} = \left| \frac{x_{i_1,\lim} - x_{i_1,cur}}{x_{i_1,\lim} - x_{i_1,in}} \right|, \qquad (1)$$

where $x_{i_1,\lim}$ is admissible limit normative value of i_1^{th} diagnostic parameter; $x_{i_1,cur}$ is value of i_1^{th} diagnostic parameter at the moment of control; $x_{i_1 in}$ is initial value of i_1^{th} diagnostic parameter (at the moment of putting into operation of new equipment or after repair), i_1 is number of diagnostic parameter.

We perform the reduction of the circuit by the following expressions. For serial part of the circuit (**Figure 1**) the coefficient of total residual resource is found by the expression:

$$k_{tot.resid.res.} = \prod_{\tau=1}^{\nu} k_{\tau}^{p_{\tau}}, \qquad (2)$$

where k_τ is the coefficient of residual resource of PT by τ^{th} diagnostic parameter; τ is τ^{th} diagnostic parameter; ν is the amount of blocks in the serial part of the circuit of **Figure 1**, p_τ probability of control parameters deviator from maximum permissible normalized value of this parameter is found by means of the expression (3):

$$p_\tau = \frac{y_\tau}{m_2}, \qquad (3)$$

where y_τ is a number of controlled parameter deviations from admissible limiting normalized value of this parameter, which were revealed by means of τth diagnostic parameter control (τ for serial part of the circuit) from the total number of the revealed deviations of controlled parameters from admissible limiting normalized value; m_2 is total quantity of the revealed deviations of controlled diagnostic parameter from their admissible limiting normalized values.

For parallel part of the circuit the coefficient of total residual resource is found by the expression (4)

$$k_{tot.resid.res.} = 1\text{-}\sum_{j=1}^{m_1}\left[\left(1\text{-}k_{res,j}\right)p_j\right], \tag{4}$$

where $k_{res,j}$ is the coefficient of residual resource of PT by jth diagnostic parameter; j is number of jth diagnostic parameter; m_1 is a quantity of blocks(parameters) in parallel part of the circuit that is reduced.

The coefficient of total residual resource of PT is determined by the expression (5):

$$k_{res} = k_{wind.} \cdot k_{in.} \cdot k_{bush} \cdot k_{oil} \cdot k_{LTC} \cdot k_{cool} \cdot k_{tank}, \tag{5}$$

where k_{wind}, k_{in}, k_{bush}, k_{oil}, k_{LIC}, k_{cool}, k_{tank} are known at the moment of calculation values of the coefficient of residual resource: of the windings, of the insulation, of bushings, of the oil, of LTC, of system of the cooling, of tank of the transformer, by the elements of the transformer, correspondingly.

For the creation of mathematical model of residual resource coefficient of the transformer parameters were used, by each of these parameters the conclusion regarding the state of the transformer can be made. But none of these parameters completely characterizes technical state of the transformer, it only shows certain changes of technical state of power transformer.

Mathematical model of residual resource coefficient of the transformer was created by means of MatLab. Using this model, it is possible to edit the already created (5) probabilistic sample of teaching data. These data help to obtain analytical dependence of residual resource coefficient of the transformer on diagnostic parameters in the form of the polynomial. For seven input parameters of the model, that randomly changed from 0 to 1, the coefficient of total residual resource of the transformer (5) was determined, where input parameters of the model were reduced to relative units of their deviation from the norm.

By means of Anfis Editor using hybrid teaching algorithm and applying Sugeno algorithm of neuro-fuzzy conclusion neuro-fuzzy model of residual resource coefficient of the transformer (using subclusterization method) was obtained.

Figure 2 contains the copy of screen saver in Matlab environment where the structure of the obtained neural network is shown.

For each input variable of neuro-model four linguistic terms with Gaussian membership functions were used:

$$k_{res.i_1} = f(x_{i_1}; \sigma_{i_1}; c_{i_1}) = e^{-\frac{\left(x_{i_1} - c_{i_1}\right)^2}{2 \cdot \sigma_{i_1}^{\ 2}}}, \tag{6}$$

where δ_{i_1} and c_{i_1} are numerical parameter; $\delta_{i_1}^2$ in probability theory it is called dispersion of the distribution(14), and the second parameter c_{i_1} is mathematic expectation; i_1 is input parameter of neuro-fuzzy model, that corresponds to diagnostic parameter (i_1 = 1, 2, 3, 4, 5, 6, 7), x_i is value of i_1^{th} input parameter of the model: $x_1 - k_{wind}$, $x_2 - k_{in}$, $x_3 - k_{bush}$, $x_4 - k_{oil}$, $x_5 - k_{LTC}$, $x_6 - k_{cool}$, $x_7 - k_{tank}$.

These are such terms as: "normal" values of diagnostic parameter, "minor deviations" of diagnostic parameter value, "prefault" values of diagnostic parameter, "emergency" value of diagnostic parameter.

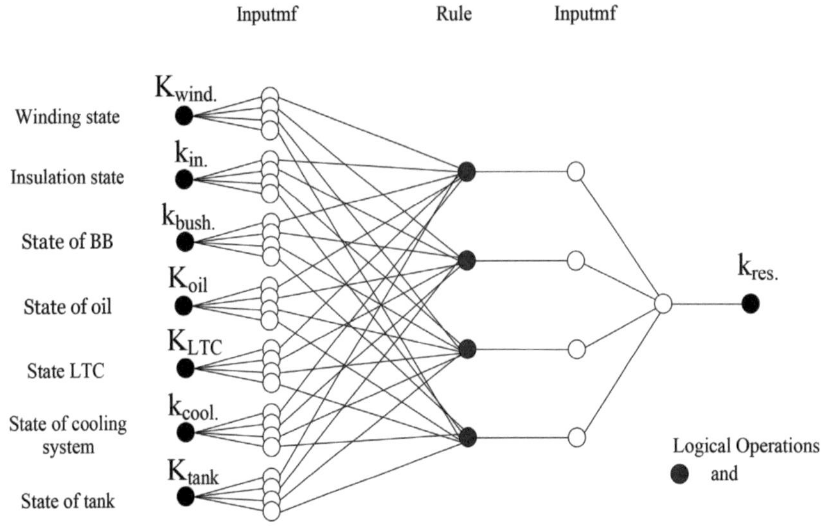

Figure 2.
Structure of Anfis-network of the transformer.

For determining the value of total residual coefficient neuro-fuzzy non-linear autoregressive model of the total residual resource coefficient of the transformer is used. This model establishes neuro-fuzzy non-linear transformation between the values of residual resource coefficients by diagnostic parameters and total residual resource coefficient of the transformer (7):

$$k_{tot.resid.res} = F(k_{wind.} \cdot k_{in.} \cdot k_{bush} \cdot k_{oil} \cdot k_{LTC} \cdot k_{cool} \cdot k_{tank}), \quad (7)$$

where F is neuro-fuzzy functional transformation.

For determination of the value of total residual resource coefficient of the transformer, we use Takagi-Sugeno model of logic conclusion.

Mathematical model of total residual resource coefficient is the system of logic equations (8).

$$\begin{cases} IF\ k_{wind.} \in \text{"normal"}\ AND\ k_{in.} \in \text{"normal"}\ AND\ k_{вush.} \in \text{"normal"} \\ AND\ k_{oil} \in \text{"normal"}\ AND\ k_{LTC} \in \text{"normal"}\ AND\ k_{cool.} \in \text{"normal"} \\ AND\ k_{tank} \in \text{"normal"}\ THEN \\ k_{tot.resid.res} = a_{1\ 1} \cdot k_{wind.} + a_{1\ 2} \cdot k_{in.} + a_{1\ 3} \cdot k_{bush.} + a_{1\ 4} \cdot k_{oil} + a_{1\ 5} \cdot k_{LTC} + a_{1\ 6} \cdot k_{cool.} + a_{1\ 7} \cdot k_{tank} + c_1 \\ IF\ k_{wind} \in \text{"minordeviations}\ AND\ k_{in} \in \text{"minor deviation"} \\ AND\ k_{вв} \in \text{"minor deviation"}\ AND\ k_{oil} \in \text{"minor deviation"} \\ AND\ k_{LTC} \in \text{"minor deviation"}\ AND\ k_{cool} \in \text{"minor deviation"} \\ AND\ k_{tank} \in \text{"minor deviation"}\ THEN \\ k_{tot.resid.res} = a_{21} \cdot k_{wind.} + a_{22} \cdot k_{in.} + a_{23} \cdot k_{bush.} + a_{24} \cdot k_{oil} + a_{25} \cdot k_{LTC} + a_{26} \cdot k_{cool.} + a_{27} \cdot k_{tank} + c_2 \\ IF\ k_{wind} \in \text{"prefault"}\ AND\ k_{in} \in \text{"prefault"}\ AND\ k_{вв} \in \text{"prefault"} \\ AND\ k_{oil} \in \text{"prefault"}\ AND\ k_{LTC} \in \text{"prefault"}\ AND\ k_{cool} \in \text{"prefault"} \\ AND\ k_{tank} \in \text{"prefault"}\ THEN \\ k_{tot.resid.res} = a_{3\ 1} \cdot k_{wind.} + a_{32} \cdot k_{in.} + a_{33} \cdot k_{bush.} + a_{34} \cdot k_{oil} + a_{35} \cdot k_{LTC} + a_{36} \cdot k_{cool.} + a_{37} \cdot k_{tank} + c_3 \\ IF\ k_{wind} \in \text{"emergency"}\ AND\ k_{in} \in \text{"emergency"}\ AND\ k_{вв} \in \text{"emergency"}\ AND\ k_{oil} \in \text{"emergency"} \\ AND\ k_{LTC} \in \text{"emergency"}\ AND\ k_{cool} \in \text{"emergency"}\ AND\ k_{tank} \in \text{"emergency"}\ \ THEN \\ k_{tot.resid.res} = a_{4\ 1} \cdot k_{wind.} + a_{4\ 2} \cdot k_{in.} + a_{4\ 3} \cdot k_{bush.} + a_{4\ 4} \cdot k_{oil} + a_{4\ 5} \cdot k_{LTC} + a_{4\ 6} \cdot k_{cool.} + a_{4\ 7} \cdot k_{tank} + c_4 \\ \ldots \end{cases} \quad (8)$$

Output of the model total $k_{tot.resid.res.}$ is found as weighted sum of conclusions (8) of rules base written in the form the system of logic equations:

$$k_{tot.resid.res.} = \sum_{j2=1}^{m3} w_{j2} \begin{pmatrix} a_{j2\,1} \cdot k_{wind.} + a_{j2\,2} \cdot k_{in.} + a_{j2\,3} \cdot k_{Bush.} + a_{j2\,4} \cdot k_{oil} + \\ + a_{j2\,5} \cdot k_{LTC} + a_{j2\,6} \cdot k_{cool.} + a_{j2\,7} \cdot k_{tank} + c_{j2} \end{pmatrix}, \quad (9)$$

where $0 \leq w_{j_2} \leq 1$ is the degree of execution (weight) of the j_2th rule that is determined by the correspondence of real changes of diagnostic parameters of the transformer.

ANFIS is the simplest network of direct propagation that contains adaptive nodes, using the teaching rules the parameters of these nodes are arranged to minimize the error between the real output of the model $k_{tot.resid.mod.}$ and real total residual resource coefficient $k_{tot.resid.res}$ of the transformer

$$\delta = \sqrt{\frac{1}{N_1} \sum\nolimits_{k_3=0}^{N_1-1} \left(k_{tot.resid.res.modk3} - k_{tot.resid.res.k3}\right)^2} \rightarrow \min, \quad (10)$$

where N is number of rows in teaching sample; k_3 is the number of the row in teaching sample, starting from the row with consecutive number "0".

Taking into account the iterative computation experiments carried out the vector of membership functions parameters is determined in **Table 2**.

Parameters	Input parameters of the model	Name of the term	Number of the rule	Parameters of membership function	
				σ	C
Winding state	$K_{wind.}$	Normal	1	0.3825	0.7944
		Minor deviation	2	0.479	0.5197
		Prefault	3	0.4903	0.5668
		Emergency	4	0.4	0.1697
Insulation state	$k_{in.}$	Normal	1	0.3653	0.8698
		Minor deviation	2	0.4642	0.6104
		Prefault	3	0.5102	0.5267
		Emergency	4	0.3949	0.1742
State of BB	$k_{bush.}$	Normal	1	0.3202	0.9221
		Minor deviation	2	0.3419	0.7649
		Prefault	3	0.4914	0.5376
		Emergency	4	0.4032	0.1925
State of oil	K_{oil}	Normal	1	0.4369	0.9273
		Minor deviation	2	0.3404	0.9674

Parameters	Input parameters of the model	Name of the term	Number of the rule	Parameters of membership function	
				σ	C
		Prefault	3	0.412	0.599
		Emergency	4	0.4031	0.2057
State LTC	K_{LTC}	Normal	1	0.3984	0.973
		Minor deviation	2	0.3316	0.963
		Prefault	3	0.4468	0.5881
		Emergency	4	0.4428	0.2349
State of cooling system	$k_{cool.}$	Normal	1	0.3439	1153
		Minor deviation	2	0.3507	0.9706
		Prefault	3	0.437	0.597
		Emergency	4	0.4263	0.2397
State of tank	K_{tank}	Normal	1	0.3454	0.9506
		Minor deviation	2	0.3801	1017
		Prefault	3	0.4582	0.6273
		Emergency	2	0.5451	0.564

Table 2.
Parameters of membership function.

It is seen from **Figure 2** that in the process of formation of the structure of neuro fuzzy model of the transformer seven inputs and one output of this model were set. Each of seven inputs has four terms. That is, each set of possible values of input parameters of the model is conventionally divided into four subsets: "normal" values of input parameter, "miner deviations" of the values of input parameter, and "prefault" values of input parameter, "emergency" values of input parameter. Membership degree of each value of input parameter to corresponding set of values is determined by Gaussian membership function. The model is intended for determining the numerical value of total residual resource coefficient of the transformer, that is why it has one output. This numerical value is found by means of solution of linear equation, that describes the dependence of the coefficient of total residual resource of the transformer on input parameters.

The obtained neuro-fuzzy model allows to determine the value of total residual resource coefficient of the transformer depending on the values of input parameters residual resources coefficients by each of controlled diagnostic parameters. The error of PPCT mathematical model changes from +0,004 relative units, if PPCT equals 0, to −0,032, when PPCT equals 1.

Taking into account the data of the **Table 1** and **Table 2** and (9) we obtain mathematical model of the coefficient of total residual resource in the form:

$$\begin{cases} \text{IF}\, k_{wind.} \in \text{"normal"}\, \text{AND}\, k_{in} \in \text{"normal"}\, \text{AND}\, k_{BB} \in \text{"normal"} \\ \text{AND}\, k_{oil} \in \text{"normal"}\, \text{AND}\, k_{LTC} \in \text{"normal"}\, \text{AND}\, k_{cool} \in \text{"normal"} \\ \text{AND}\, k_{tank} \in \text{"normal"}\ \text{THEN} \\ k_{tot.resid.res} = 0,6166 \cdot k_{wind} + 0,4125 \cdot k_{in} + 0,4618 \cdot k_{BB} + 1,83 \cdot k_{oil} + 1,804 \cdot k_{LTC} + \\ +0,0462 \cdot k_{cool.} + 1,96 \cdot k_{tank} \text{-} 5,377 \\ \text{IF}\, k_{tank} \in \text{"minor deviation"}\, \text{AND}\, k_{in.} \in \text{"minor deviation"} \\ \text{AND}\, k_{Bush} \in \text{"minor deviation"}\, \text{AND}\, k_{oil}. \in \text{"minor deviation"} \\ \text{AND}\, k_{LTC} \in \text{"minor deviation"}\, \text{AND}\, k_{cool.} \in \text{"minor deviation"} \\ \text{AND}\, k_{tank} \in \text{"minor deviation"}\ \text{THEN} \\ k_{tot.resid.res} = \text{-}0,0393 \cdot k_{wind.} + 0,2609 \times k_{in.} + 0,1086 \times k_{Bush.} \text{-} 0,37 \cdot k_{oil} \text{-} 0,1459 \cdot k_{LTC} \text{-} \\ \text{-}0,02387 \cdot k_{cool.} \text{-} 0,05863 \cdot k_{tank} + 0,1288 \\ \text{IF}\, k_{wind} \in \text{"prefault"}\, \text{AND}\, k_{in} \in \text{"prefault"}\, \text{AND}\, k_{Bush.} \in \text{"prefault"} \\ \text{AND}\, k_{oil} \in \text{"prefault"}\, \text{AND}\, k_{LTC} \in \text{"prefalt"}\, \text{AND}\, k_{cool.} \in \text{"prefault"} \\ \text{AND}\, k_{tank} \in \text{"prefault"}\ \text{THEN} \\ k_{tot.resid.res} = \text{-}0,2165 \cdot k_{wind.} \text{-} 0,3714 \cdot k_{in.} \text{-} 0,4678 \times k_{Bush.} \text{-} 0,514 \cdot k_{oil} \text{-} 0,882 \cdot k_{LTC} \text{-} \\ \text{-}0,5302 \cdot k_{cool.} \text{-} 1,406 \cdot k_{tank} + 3,88 \\ \text{IF}\, k_{wind.} \in \text{"emergency"}\, \text{AND}\, k_{in} \in \text{"emergency"}\, \text{AND}\, k_{BB} \in \text{"emergency"}\, \text{AND}\, k_{oil} \in \text{"emergency"} \\ \text{AND}\, k_{LTC} \in \text{"emergency"}\, \text{AND}\, k_{cool.} \in \text{"emergency"}\, \text{AND}\, k_{tank} \in \text{"emergency"}\ \text{THEN} \\ k_{tot.resid.res} = 0,03166 \cdot k_{wind.} \text{-} 0,06144 \cdot k_{in.} \text{-} 0,387 \cdot k_{Bush.} + 0,06 \cdot k_{oil} + 0,3199 \cdot k_{LTC} \text{-} \\ \text{-}0,026 \cdot k_{cool.} \text{-} 0,006 \cdot k_{tank} + 0,003 \\ \ldots \end{cases} \tag{11}$$

The obtained neuro-fuzzy model allows to determine the value of total residual resource coefficient of the transformer depending on the values of input parameters residual resources coefficients by each of controlled diagnostic parameters. The error of PPCT mathematical model changes from +0.004 relative units, if PPCT equals 0, to −0.032, when PPCT equals 1.

Despite the complexity of dependences, mathematical model of residual resource coefficient of the transformer (11) may be used for programming neuro-fuzzy controller in order to create the device for on-line determination of transformer state by means of analysis of residual resource coefficient of the transformer value.

6. Account of the forecast current value of residual resource of the transformers in the process of control of EES modes

It is known that in the process of operation, energy enterprise plans to remove out of service the equipment in the overhaul, cost of is forecast. Removal of the transformer into overhaul in a planned number of years (T_{WF}) of trouble-free operation (12 years) provides certain list of works and their expected cost $B_{oh\ pl.}$ For instance, for 330/110 kV transformers of 125–250 MVA power the cost (B) of such repair is 770–11,550 $. We propose to assume that removal out of service the transformers into current repair requires unscheduled cost.

The cost of repair may increase by the value ΔB_1, replacement of damaged blocks of the transformer and additional work, connected with the replacement. These costs are not provided in case of “typical” planned overhaul

$$\Delta B_1 = \sum_{i=1}^{n} \left(B_i \cdot e^{\gamma_i \cdot k_{res.i}^{\beta_i}} \right), \tag{12}$$

where B_i is the cost of replacement of *i*th damaged block of the transformer and additional work, connected with this replacement, n is a number of damaged blocks that require unscheduled replacement; $k_{res.i}^{\beta_i}$ is residual resource coefficient of *i*th block that requires unscheduled replacement; γ and β are coefficients, that characterize the impact of residual resource coefficient on the expected cost of unscheduled repair or replacement of *i*th block of the transformer (is determined by means of processing of statistic data).

Repair cost may increase by the cost of ΔB_2 (as compared with expected) in case of enlarged current (instead of planned overhaul) repair of the transformer, that did not operate for planned number of years:

$$\Delta B_{2_j} = \left(1\text{-}e^{\alpha_j (T_j\text{-}1)} \right) \cdot B_{OH}, \tag{13}$$

where j is a number of the transformer, T_j is time, the *i*th transformer functioned after putting into operation or after the last overhaul (enlarged current) repaint to the moment of mode control, λ is the coefficient, that characterizes the intensity of cost growth ΔB_2 that depends on the construction of the transformer, conductions and operation mode (is determined experimentally), B_{OH} is the cost of transformer overhaul.

It should be noted that removal the transformer out of service takes place not only as a result of relay protection, emergencies control automation operation but also by a person responsible for safety operation by the results of control of diagnostic parameter, values of which sometimes only approaches to limiting values.

Within the context of creation of modern Smart Grids and to provide safe, reliable, quality and economic efficient operation of EES it is necessary to perform the control over active power overflow to realize by means of the transformer, performing reliable and information archons on the mode. That is why, we suggest to take into account the coefficient of regulating transformer limitation:

$$k_{wind,j} = \left(1\text{-}k_{res,j} \right) \cdot B_{cq,j}, \tag{14}$$

where B_{cq} is the coefficient of repair cost value growth of the jth transformer.

$$B_{cq,j} = \frac{\Delta B_{1,j} + \Delta B_{2,j}}{\Delta B_{1,j} + B_{pl.,j}}. \tag{15}$$

As the example, we will consider 23 nodes 230/138 kV test circuit (**Figure 3**). In branches 11–9, 12–14, 12–9, 11–4, and 3–7 transformers ATDCTN-63000/230/138, ATDCTN-100000/230/138 and ATDCTN-125000/230/138 are installed. Initial node loads, complex transformation ratios and corresponding transformers LTC positions (number of taps) are given in **Tables 3** and **4**.

Knowing the circuit and normal node parameters we define transformation ratios.

$$k_{a.opt} = 1\text{-diag}\left(\text{Re}\left(\text{-}N_{k.bal._b} \cdot Z \cdot C_e \cdot J \right) \right) \cdot U_b^{-1} \cdot E_{bal.a}^{*} \tag{16}$$

$$k_{r.opt} = \text{-diag}\left(\text{Im}\left(\text{-}N_{k.bal._b} \cdot Z \cdot C_e \cdot J \right) \right) \cdot U_b^{-1} \cdot E_{bal_r}^{*} \tag{17}$$

where $N_{k.bal.b}$ is the second matrix of branches connection in contour balanced transformation ratios; Z diagonal matrix of complex branches resistances; C_e is the

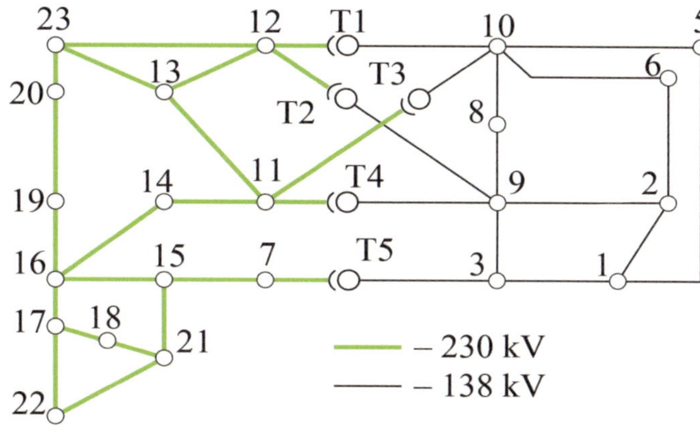

Figure 3.
Scheme of grid 230/138 kV for 23 nodes.

matrix of currents distribution coefficients for economic mode of electric network (corresponds to minimal losses of electric energy); J vector-column of currents in nodes; U_b is the voltage of basic node; $E^*_{bal.a}$ and $E^*_{bal.r}$ are balancing is electric moving in relative units (EMF) in relative units (active and reactive components).

Taking in the consideration the discrete character of LTC switching's, errors of instrument transforms, errors of data transmission channels and recommendations [20] we assume that non-sensitivity zone of active power losses may be considered as regulating actions on LTC of the transformer—to be 3% [21].

As initial conditions we assume that in accordance with load graph LTC of transformers 9–11, 9–12, 11–10, 12–10 have transformation ratios 0.6413 (14 tap), 0.6347 (14 tap), 0.6397 (14 tap), 06446 (14 tap).

It should be noted that further changes of operation mode were realized at admissible voltage deviations ±5%, from nominal voltage U_{nom}.

Regulation of the transformer 7–3 is inexpedient on conditions of the usage of the given technique of determination of transformation ratios.

We define the losses of active and reactive power in the branches of the circuit at current transformation ratios (**Table 5**).

$$\Delta S_{\Sigma br} = \Delta P_{\Sigma br} + j\Delta Q_{\Sigma br} = 3 \cdot \sum_{j=1}^{m} \Delta S_{br,j}, \tag{18}$$

where $\Delta S_{br} = \mathrm{diag}\,(\Delta U_{br}) \cdot \hat{I}_{br}$ is vector-column of complete power losses in the branches of the circuit. E_{br} is vector column of the current in branches, m is a number of the branch in the circuit, $\Delta U_{br} = M_{\Sigma} \cdot U_{node}$ is vector-column of phase voltages in the nodes, ΔP_{br}, ΔQ_{br} is vector-column of active and reactive power losses in the branches of the circuit(correspondingly).

We define transformation ratios (16–17) and position of LTC on condition of minimal amount of switchings (in order to maintain switching resources of LTC) to provide minimal losses of active power in branches of the circuit of **Table 6**.

As a result of realization of control actions, mode optimization power losses were reduced from ΔS_1 = 4.49 + j29.05 (MVA) for mode (**Table 5**) to ΔS_2 = 4.42 + j28.69 (MVA). Thus, the effect of realization of transformer LTC switchings is ΔS_1-ΔS_2 = 0.07 + j0.36 (MVA). The transition from the current to mode may be performed by switching the LTC of the transformer, installed in the branch 9–12 from 13 tap to 14 and carry out transformer regulation changing position of LTC from 14 tap to 15. We will consider the transition to another stage of

Branches		R (Ohm)	X (Ohm)	k_{active}	$k_{reactive}$
№ of beginning	№ of the end				
11	10	0.6	27	0.6487	0
12	9	0.37	9.28	0.6498	0
11	9	0.3	13	0.6479	0
7	3	0.21	11.53	0.65	0
1	2	0.4951	26.471	1	0
1	3	10.398	40.221	1	0
1	5	41.516	16.092	1	0
2	4	62.464	24.129	1	0
2	6	94.649	36.565	1	0
3	9	9.882	20.962	1	0
4	9	51.038	19.749	1	0
5	10	4.342	16.816	1	0
8	9	8.82	15.124	1	0
8	10	2.067	2.145	1	0
11	23	5.207	22.793	1	0
11	14	28.566	22.112	1	0
12	23	5.207	25.18	1	0
12	13	65.596	51.101	1	0
23	13	58.719	45.759	1	0
14	16	2.645	20.578	1	0
15	16	2.338	8.404	1	0
16	17	17.457	13.701	1	0
16	19	3117	11.206	1	0
17	18	0.9522	76.176	1	0
17	22	71.415	55.704	1	0
21	22	46.023	35.866	1	0
7	15	5.68	24.865	1	0
21	18	0.873	6.851	1	0
21	15	1.666	12.96	1	0
19	20	1.349	10.474	1	0
20	13	0.741	5.713	1	0
12	10	0.31	14	0.6524	0
10	6	26.471	11.522	1	0

Table 3.
Information of circuit branches.

daily load graph(load increase), its parameters are given in **Table 7**, and optimized transformation ratios and corresponding mode parameters - in **Table 8**.

As a result of performing control actions, mode optimization we succeeded in decreasing power losses from ΔS_1 = 61.85 + j412.73 (MVA) for the mode (**Table 7**)

№ of the node	U (kV)	Phase (grad)	P_{load} (MW)	Q_{load} (MVAr)	P_{gen} (MW)	Q_{gen} (MW)
1	136.34	−5.73	108	22	182	30
2	135.54	−6.27	187.15	76	172.9	30
3	141.22	−0.94	176.4	36.26	0	0
4	136	−8.2	74	15	0	0
5	137	−8.11	68.16	13.44	0	0
6	136.36	−10.3	129.2	25.27	0	0
7	218.84	2.99	19.4	1.94	0	0
8	142.57	−7.72	169.29	34.65	131.55	131.75
9	141.55	−5.6	275	66	0	0
10	141.64	−7.29	191.1	49	0	0
11	222.36	−3.12	40	10	0	0
12	219.92	−4.12	54.88	17.64	0	0
13	236.15	7.89	0	0	495	150
14	228.34	1.56	184.3	37.05	0	101.39
15	233.22	11.21	304.32	61.44	235.2	51.32
16	233.45	10	100	20	185	80
17	237.79	15.28	35.64	13.86	0	0
18	241.5	16.92	323.01	65.96	417.1	176.19
19	231.47	7.6	177.38	36.26	0	0
20	233.81	7.31	128	26	0	0
21	241.5	17.83	0	0	425.7	146
22	241.5	25.95	0	0	420	−3.69
23	234.6	0	265	54	417.85	281.37

Table 4.
Information of the circuit nodes.

to ΔS_2 = 61.80 + j412.68 (MVA). Thus the effect of LTC transformer switchings is ΔS_1-ΔS_2 = 0.05 + j0.5 (MVA).

If as a result of determining the coefficient of regulating effect limitation for circuit transformers (**Figure 3**) the following values are obtained: $k_{wind,9–11}$ = 0.85, $k_{wind,12–9}$ = 0.4, $k_{wind,11–10}$ = 0.3 a $k_{wind,12–10}$ = 0,2, then expected quasi-decrease of losses, taking into account these coefficient will be defined.

Control actions are performed by the transformer, installed in the branch 9–11, namely, we change position of LTC with 14 tap on 15, in this case, the expected losses of active power are $\Delta P_{9–11}$ = 61.82 (MW). We find the decrease of active power losses ΔP_Σ - $\Delta P_{9–11}$ = 61.85–61.82 = 0.03 (MW), however, having taken into account the coefficient of regulating effect limitation, losses decrease change $\delta P_{quasi.9–11}$ = (ΔP_Σ - $\Delta P_{9–11}$)•$k_{wind,9–11}$ = 0.0255 (MW). New quasi-losses $\Delta P_{quasi.9–11}$ = $\Delta P_{9–11}$ + $\delta P_{quasi.9–11}$ = 61.82 + 0.0255 = 61.8455 (MW). Results of the calculation of other transformers are given in **Table 9**.

We define mode parameters for the circuit with quasi-resistances from **Table 11** and corrected transformation ratios from **Table 12**.

We find losses of active power in the branch, that contains the transformer, as function the element of vector-column of complete power losses in the branches of the circuit by the expression

P-ters*	№ of nodes																					
	1	**2**	**3**	**4**	**5**	**6**	**7**	**8**	**9**	**10**	**11**	**12**	**13**	**14**	**15**	**16**	**17**	**18**	**19**	**20**	**21**	**22**	**23**
P_{load}, MW	19.44	37.43	38.8	17.02	13.63	24.55	3.49	27.08	60.5	38.22	8.4	10.43	0	29.49	39.6	20	8.55	80.75	35.47	26.88	0	0	53
Q_{load}, MVAr	3.96	15.2	7.97	3.45	2.69	4.8	0.35	5.54	14.52	9.8	2.1	3.35	0	5.93	8	4	3.33	16.49	7.25	5.46	0	0	10.8
ΔP_{Σ}, MW	4.49																						
ΔQ_{Σ}, MVar	29.05																						
11–10	—	—	—	—	—	—	—	—	—	0.6437		—	—	—	—	—	—	—	—	—	—	—	—
№ tap	—	—	—	—	—	—	—	—	—	14		—	—	—	—	—	—	—	—	—	—	—	—
12–10	—	—	—	—	—	—	—	—	—		0.6446		—	—	—	—	—	—	—	—	—	—	—
№ tap	—	—	—	—	—	—	—	—	—		14		—	—	—	—	—	—	—	—	—	—	—
9–11	—	—	—	—	—	—	—	—	0.6413			—	—	—	—	—	—	—	—	—	—	—	—
№ tap	—	—	—	—	—	—	—	—	14			—	—	—	—	—	—	—	—	—	—	—	—
9–12	—	—	—	—	—	—	—	—	0.6397				—	—	—	—	—	—	—	—	—	—	—
№ tap	—	—	—	—	—	—	—	—	13				—	—	—	—	—	—	—	—	—	—	—
9–12	—	—	0.65					—	—	—	—	—	—	—	—	—	—	—	—	—	—	—	—
№ tap	—	—	14					—	—	—	—	—	—	—	—	—	—	—	—	—	—	—	—

*Parameters.

Table 5.
Parameters of the current mode.

P-ters	№ of nodes																					
	1	2	3	4	5	6	7	8	9	10	11	12	13	14	15	16	17	18	19	20	21	22	23
P_{load}, MW	19.44	37.43	38.8	17.02	13.63	24.55	3.49	27.08	60.5	38.22	8.4	10.43	0	29.49	39.6	20	8.55	80.75	35.47	26.88	0	0	53
Q_{load},. MVAr	3.96	15.2	7.97	3.45	2.69	4.8	0.35	5.54	14.52	9.8	2.1	3.35	0	5.93	8	4	3.33	16.49	7.25	5.46	0	0	10.8
ΔP_{Σ},. MW	4.42																						
ΔQ_{Σ},. MVar	28.69																						
11–10	—	—	—	—	—	—	—	—	—	0.6513		—	—	—	—	—	—	—	—	—	—	—	—
№ tap	—	—	—	—	—	—	—	—	—	14		—	—	—	—	—	—	—	—	—	—	—	—
12–10	—	—	—	—	—	—	—	—	—	0.6542			—	—	—	—	—	—	—	—	—	—	—
№ tap	—	—	—	—	—	—	—	—	—	15			—	—	—	—	—	—	—	—	—	—	—
9–11	—	—	—	—	—	—	—	—	0.6507			—	—	—	—	—	—	—	—	—	—	—	—
№ tap	—	—	—	—	—	—	—	—	14			—	—	—	—	—	—	—	—	—	—	—	—
9–12	—	—	—	—	—	—	—	—	0.6521				—	—	—	—	—	—	—	—	—	—	—
№ tap	—	—	—	—	—	—	—	—	14				—	—	—	—	—	—	—	—	—	—	—
9–12	—	—	0.65					—	—	—	—	—	—	—	—	—	—	—	—	—	—	—	—
№ tap	—	—	14					—	—	—	—	—	—	—	—	—	—	—	—	—	—	—	—

Table 6.
Transformation ratios for the current mode.

P-ters	№ of nodes																					
	1	2	3	4	5	6	7	8	9	10	11	12	13	14	15	16	17	18	19	20	21	22	23
P_{load}, MW	108	187.1	176.4	74	68.16	129.2	19.4	169.2	275	39.1	40	54.88	0	184.3	304.3	100	35.64	323.0	177.3	128	0	0	265
Q_{load}, MVAr	22	76	36.26	15	13.44	25.27	1.94	34.65	66	10	10	17.64	0	37.05	61.44	20	13.86	65.96	36.26	26	0	0	54
ΔP_{Σ}, MW												61.85											
ΔQ_{Σ}, MVar												412.73											
11–10	—	—	—	—	—	—	—	—	—	0.6513		—	—	—	—	—	—	—	—	—	—	—	—
№ tap	—	—	—	—	—	—	—	—	—	14		—	—	—	—	—	—	—	—	—	—	—	—
12–10	—	—	—	—	—	—	—	—	—		0.6542		—	—	—	—	—	—	—	—	—	—	—
№ tap	—	—	—	—	—	—	—	—	—		15		—	—	—	—	—	—	—	—	—	—	—
9–11	—	—	—	—	—	—	—	—		0.6507		—	—	—	—	—	—	—	—	—	—	—	—
№ tap	—	—	—	—	—	—	—	—		14		—	—	—	—	—	—	—	—	—	—	—	—
9–12	—	—	—	—	—	—	—	—			0.6521		—	—	—	—	—	—	—	—	—	—	—
№ tap	—	—	—	—	—	—	—	—			14		—	—	—	—	—	—	—	—	—	—	—
9–12	—	—			0.65			—	—	—	—	—	—	—	—	—	—	—	—	—	—	—	—
№ tap	—	—			14			—	—	—	—	—	—	—	—	—	—	—	—	—	—	—	—

Table 7.
Parameters of the mode after load change.

P-ters	№ of nodes																					
	1	2	3	4	5	6	7	8	9	10	11	12	13	14	15	16	17	18	19	20	21	22	23
P_{load}, MW	108	187.1	176.4	74	68.16	129.2	19.4	169.2	275	39.1	40	54.88	0	184.3	304.3	100	35.64	323.0	177.3	128	0	0	265
Q_{load}, MVAr	22	76	36.26	15	13.44	25.27	1.94	34.65	66	10	10	17.64	0	37.05	61.44	20	13.86	65.96	36.26	26	0	0	54
ΔP_{Σ}, MW	61.80																						
ΔQ_{Σ}, MVar	412.68																						
11–10	—	—	—	—	—	—	—	—	—	0,665		—	—	—	—	—	—	—	—	—	—	—	—
№ tap	—	—	—	—	—	—	—	—	—	16		—	—	—	—	—	—	—	—	—	—	—	—
12–10	—	—	—	—	—	—	—	—	—	0.651			—	—	—	—	—	—	—	—	—	—	—
№ tap	—	—	—	—	—	—	—	—	—	14			—	—	—	—	—	—	—	—	—	—	—
9–11	—	—	—	—	—	—	—	—	0.6753			—	—	—	—	—	—	—	—	—	—	—	—
№ tap	—	—	—	—	—	—	—	—	17			—	—	—	—	—	—	—	—	—	—	—	—
9–12	—	—	—	—	—	—	—	—	0.659				—	—	—	—	—	—	—	—	—	—	—
№ tap	—	—	—	—	—	—	—	—	15				—	—	—	—	—	—	—	—	—	—	—
7–3	—	—	0.65					—	—	—	—	—	—	—	—	—	—	—	—	—	—	—	—
№ tap	—	—	14					—	—	—	—	—	—	—	—	—	—	—	—	—	—	—	—

Table 8.
Parameters of mode without taking into account technical state of the transformers.

Transf.	$\frac{K_{tr.cur.}}{N_{cur.}}$	$\frac{K_{tr.opt.}}{N_{opt.}}$	$K_{wind.\ j}$	$\Delta P_{tr.j}$ (MW)	ΔP_{Σ} - $\Delta P_{tr.j}$ (MW)	$\delta P_{quasi.j}$ (MW)	$\Delta P_{quasi\ j.}$ (MW)
9–11	0.6507 / **14**	0.659 / **15**	0.85	**61.82**	0.03	0.0255	61.8455
12–9	0.6521 / **14**	0.6419 / **13**	0.46	61.83	0.02	0.0092	**61.8392**
11–10	0.6513 / **14**	0.665 / **16**	0.34	61.835	0.015	0.0051	61.8401
12–10	0.6542 / **15**	0.651 / **14**	0.25	61.84	0.01	0.0025	61.8425

Table 9.
Results of limiting effect coefficients calculation for circuit transformers.

$$\Delta P_{\alpha} = \text{Re}\left(\Delta S_{\alpha}\right), \tag{19}$$

where $\Delta S_{\alpha} = \Delta U_{\alpha} \bullet I_{\alpha}$ is element of vector-column of power losses in the branches, that contain transformers, ΔU_{α} is kth element of vector-column of phase voltages drop in the branches, and I_{α} is the current of the branches with transformers couplings, α is the number of row, that correspond to the branch with transformer couplings in vector-column ΔS_{br}.

The value of quasi resistance in kth-branch:

$$Z_{\alpha} = \frac{\Delta S_{\alpha}}{\hat{I}_{\alpha}^{2}}, \tag{20}$$

where $\alpha = k + \beta$, where k – is the number of the row of the first branch, that contains the transformer, β is the coefficient of the change of consecutive number of branch, that contains the transformer, it changes in the range from 0 to$(\psi - 1)$, ψ is the number of branches, containing transformers.

Applying this algorithm, according to (20), quasi-resistances of the branches, containing transformers are found. The results of the calculations are given in **Table 10**. The aim of control is provider of minimum of all system active power losses that is determined by the expression

$$\Delta F = \sum_{i=1}^{n} \Delta P_{i} \rightarrow \min. \tag{21}$$

If $\Delta F_{\min.} = \Delta P_{\min.}$ is the minimum value of the efficiency function (active power losses), $\Delta F_{cur.} = \Delta P_{cur.}$ current value of efficiency function (active power

Parameters	Transformer 9–11	Transformer 12–9	Transformer 11–10	Transformer 12–10
Branch resistance, Ohm	0.3 + j13	0.37 + j9.28	0.3 + j27	0.3 + j14
Quasi resistance of the branch, Ohm	0.32 + j25.2	0.4 + j13.2	0.36 + j28.4	0.35 + j19.2

Table 10.
Quasi-resistances of transformers branches of the circuit.

P-ters	№ of nodes																					
	1	**2**	**3**	**4**	**5**	**6**	**7**	**8**	**9**	**10**	**11**	**12**	**13**	**14**	**15**	**16**	**17**	**18**	**19**	**20**	**21**	**22**	**23**
P_{load}, MW	108	187.1	176.4	74	68.16	129.2	19.4	169.2	275	39.1	40	54.88	0	184.3	304.3	100	35.64	323.0	177.3	128	0	0	265
Q_{load}, MVAr	22	76	36.26	15	13.44	25.27	1.94	34.65	66	10	10	17.64	0	37.05	61.44	20	13.86	65.96	36.26	26	0	0	54
ΔP_{Σ}, MW	62.47																						
ΔQ_{Σ}, MVar	422.16																						
11–10	—	—	—	—	—	—	—	—	—	0.665		—	—	—	—	—	—	—	—	—	—	—	—
№ tap	—	—	—	—	—	—	—	—	—	16		—	—	—	—	—	—	—	—	—	—	—	—
12–10	—	—	—	—	—	—	—	—	—		0.651		—	—	—	—	—	—	—	—	—	—	—
№ tap	—	—	—	—	—	—	—	—	—		14		—	—	—	—	—	—	—	—	—	—	—
9–11	—	—	—	—	—	—	—	—		0.659		—	—	—	—	—	—	—	—	—	—	—	—
№ tap	—	—	—	—	—	—	—	—		15		—	—	—	—	—	—	—	—	—	—	—	—
9–12	—	—	—	—	—	—	—	—		0.6419			—	—	—	—	—	—	—	—	—	—	—
№ tap	—	—	—	—	—	—	—	—		13			—	—	—	—	—	—	—	—	—	—	—
7–3	—	—			0.65			—	—	—	—	—	—	—	—	—	—	—	—	—	—	—	—
№ tap	—	—			14			—	—	—	—	—	—	—	—	—	—	—	—	—	—	—	—

Table 11.
Parameters of normal mode after loads change, taking into account technical stale of transformers.

P-ters	№ of nodes																					
	1	**2**	**3**	**4**	**5**	**6**	**7**	**8**	**9**	**10**	**11**	**12**	**13**	**14**	**15**	**16**	**17**	**18**	**19**	**20**	**21**	**22**	**23**
P_{load}, MW	108	187.1	176.4	74	68.16	129.2	19.4	169.2	275	39.1	40	54.88	0	184.3	304.3	100	35.64	323.0	177.3	128	0	0	265
Q_{load}, MVAr	22	76	36.26	15	13.44	25.27	1.94	34.65	66	10	10	17.64	0	37.05	61.44	20	13.86	65.96	36.26	26	0	0	54
ΔP_{Σ}, MW												62.46											
ΔQ_{Σ}, MVar												422.28											
11–10	—	—	—	—	—	—	—	—	—	0.664		—	—	—	—	—	—	—	—	—	—	—	—
№ tap	—	—	—	—	—	—	—	—	—	15		—	—	—	—	—	—	—	—	—	—	—	—
12–10	—	—	—	—	—	—	—	—	—		0.641		—	—	—	—	—	—	—	—	—	—	—
№ tap	—	—	—	—	—	—	—	—	—		13		—	—	—	—	—	—	—	—	—	—	—
9–11	—	—	—	—	—	—	—	—		0.6753		—	—	—	—	—	—	—	—	—	—	—	—
№ tap	—	—	—	—	—	—	—	—		17		—	—	—	—	—	—	—	—	—	—	—	—
9–12	—	—	—	—	—	—	—	—			0.6492		—	—	—	—	—	—	—	—	—	—	—
№ tap	—	—	—	—	—	—	—	—			14		—	—	—	—	—	—	—	—	—	—	—
7–3	—	—			0.65			—	—	—	—	—	—	—	—	—	—	—	—	—	—	—	—
№ tap	—	—			14			—	—	—	—	—	—	—	—	—	—	—	—	—	—	—	—

Table 12.
Parameters of mode after loads change, taking into account technical state of transformers and corrected transformation ratios.

losses), n is the total number of branches in the circuits, $k_{TT.min.}$ value of transformation ratio at which calculated losses of active power are minimal, then the dependence of active power losses change (values of efficiency function $\Delta F^*_{cur.} = \frac{\Delta F_{cur.}}{\Delta F_{min.}}$) in relative units on the values of transformation ratios $k* = \frac{k_{cur.}}{k_{tr.min.}}$ (**Figure 4**) for various transformers will be built.

Thus, the transition from the current to the mode can be realized by switching the LTC of the transformer, installed in the branch 9–12 from 13 tap to 14tap and perform regulation of the transformer in branch 12–10, changing LTC position from tap 14 to 15 tap are shown in **Figure 5**.

As a result of realization of control actions the mode will be reached by transformer switching of the branch 11–10 from 14 to 16 tap of LTC, transformer of the branch 12–10 from 15 tap to 14 tap, transformer of the branch 9–11 transformer from 14 to 15 tap and transformer of the branch 9–12 from 14 to 13 tap of LTC, respectively, how are shown in **Figure 6**.

We see that due to consideration of technical state of transformers, their ranking occurred by the measure of impact on the reduction of active power losses. To reach mode now it is more expedient to use a transformer of 9–12 branch as it during one switching of LTC from 14 tap to 15 tap reduces most active power losses.

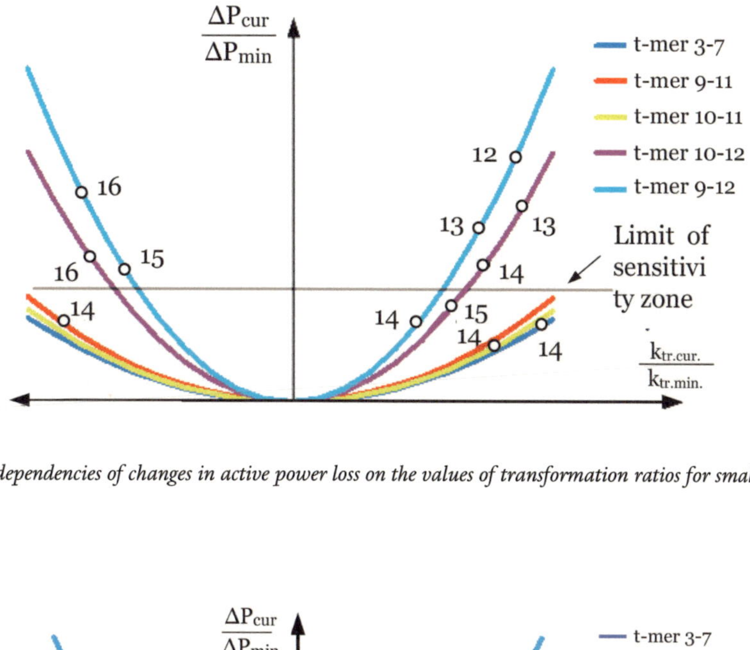

Figure 4.
Charts of dependencies of changes in active power loss on the values of transformation ratios for small loads mode.

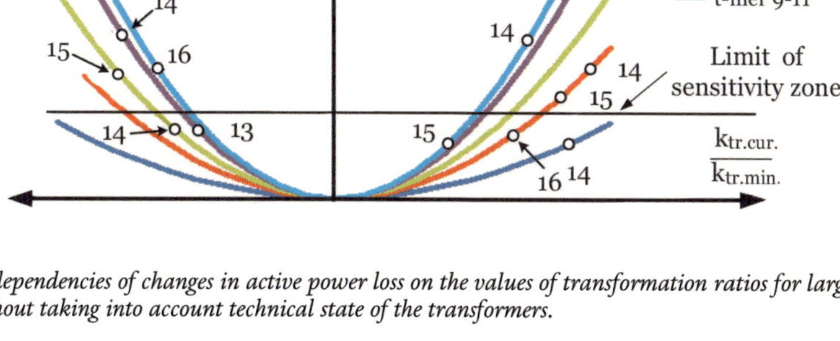

Figure 5.
Charts of dependencies of changes in active power loss on the values of transformation ratios for large loads mode, without taking into account technical state of the transformers.

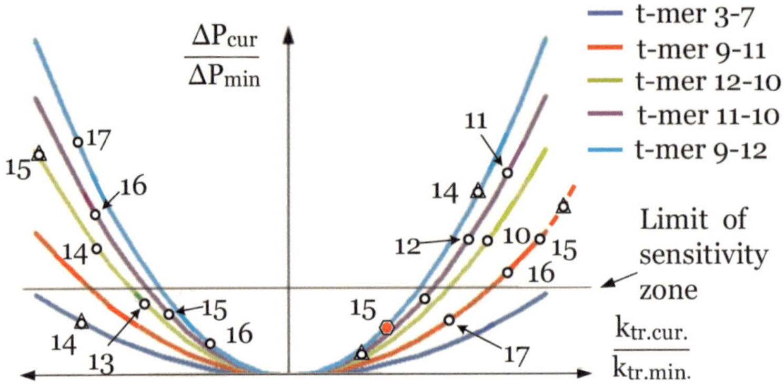

Figure 6.
Transformation ratio for large loads mode, taking into account technical state of the transformers.

7. Discussion of the results of transformation ratios of EES transformers determination, taking into account the state of transformers

The analysis of the articles showed that neuro-fuzzy logic methods are used to solve various problems of operating electric power systems, such as improving power quality [18], classifying the faults of the electrical equipment based on sequence components [19], developing of the controller of the tariff, which based on neuro-fuzzy logic for the for distributing active power between a micro network and EPC for improving energy quality [20].

The error of RRCT determination by means of the developed mathematical neuro-fuzzy model, as compared with teaching sample and to the opinion of independent experts does not exceed the error of the devices, measuring diagnostic parameters. Such results are explained by complex usage of probability theory methods neuro-fuzzy modeling and modern software Matlab. These results also confirm the information provided in the article by Moudud Ahmed, Naruttam Kumar Roy. In their article [21], it is written that the use of automatic systems for adaptive control of electric power systems (EPS) based on neuro-fuzzy modeling and based on an inference system (ANFIS) is promising method. This improves EPS performance, for example, reduces power losses. Similar positive results of using neuro-fuzzy logic are described in the article by Priyanka Ray and A.K. Sinha [22]. This article says that the use of neuro-fuzzy logic has allowed the development of a hybrid control system that provides the maximum generated electrical power of hydro, wind and solar power plants even under incomplete data on current weather conditions and power consumption.

Also, in the works [23–25] of the authors H. Suna, R. M. Velasquez; J.W.M. Lara; Dong Ling; Yao-Yu Xu; Yu Liang; Yuan Li; Ning Liuand Quan and Jun Zhang were reviewed methods of intelligent diagnostics of transformers that use fuzzy logic and in the future can be applied to improve diagnostic systems and other power equipment.

Such feature of the suggested method for determining the control actions of LTC-transformers, as account of PPCT, in the process of ES modes control, provides such advantages as reduction of the equipment damage rate, decrease of active power losses in the EPS. Due to the of the peculiarities method of determination of control action of LTC-transformers, taking into account their technical state, perspectives of developments and introduction in EPS of modern microprocessor –based systems, automatic control of LTC of transformers become possible.

As compared with the known method of voltage drop control on the branches of EPS circuits, with the method of overloads decrease of transmission lines, at the expense of redistribution of power overflows in EPS, decrease of active power losses in the process of transportation by means of LTC- transformers, the suggested method allows to select, by means of account the suggested RRCT, the transformer for EPS mode control, that would simultaneously provide the reduction of power losses and is more reliable.

Usage of quasi-resistances of circuit branches, that unlike the transformers used, in the process of calculation of nominal resistances of the branches, take into account transformers state and possible losses of utility companies due to possible damages, allows to calculate EPS mode in rise of transformers transformation ratio change and by means comparison of calculated power losses select the most efficient transformer.

The suggested peculiarity of application the method of neuro-fuzzy modeling (usage in teaching sample the model of transformer resource instead of measured values of diagnostic parameters - calculated and partially corrected by independent experts of coefficients of residual resource) enables to take into account simultaneous impact on RRCT the results of both current and periodic control.

The drawback of the suggested mathematical neuro-fuzzy model of RRCT is necessity of large data base regarding coefficient of residual resource of diagnostic parameters CRRDP (Coefficient of residual resource of the diagnostic parameter) for specific transformers. Attempt to reduce database or use the model from other similar transformer results in the increase of model error. Limitation on the usage of RRCT model is the necessity of application only on one – investigated transformer. Therefore, we need models for each transformer. The method of determination of control actions by LTC transformers does not take into account voltage limitations in nodes and current limitations in the branches of the circuit.

Further development of the given research will be realized in the development of mathematical models of other types of high voltage equipment, involved in the process of EPS modes control, damage of which areas place (**Figure 7**).

Problems of the considered research development are caused by the necessity of long lasting experiments and observations over the processes of aging and development of high voltage equipment damage, processes of EPS modes parameters change not only on computer ad mathematical models of the equipment and EPS modes and on real equipment.

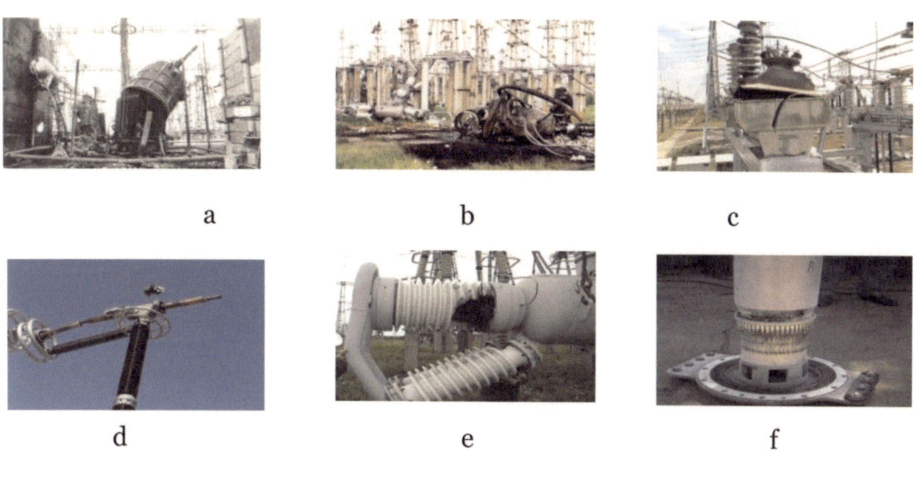

Figure 7.
Damage of high voltage equipment in EPS: (a) 750 kV shunt reactor; (b) current transformer and 750 kV air circuit breaker; (c) 33 kV voltage transformer; (d) 750 kV SF_6 circuit breaker; (e) air circuit breaker; (f) 110 kV SF_6 circuit breaker.

8. Conclusions

1. Analysis of damager ate of power transformers and methods of the EPS modes control allows to state that it is a necessary to use the results of on-line diagnostics of LTC-transformers not only to determine the expediency of further operation or repair of the equipment and for calculation transformation values (with the account of the suggested RRCT) for their usage in the process of modes control.

2. The model enables, by means of accounting, of both current and retrospective values of diagnostic parameters on RRCT and determine its current value. That is necessary for automatic and automated reliable and control of EPS modes.

3. Improved method of determination of control actions by LTC- transformers, by means of comparative analysis of the results of EES modes with quasi resistances of circuit branches, enables to soled the transformer and calculate transformation ratio that provides minimal amount of LTC switching.

Author details

Lezhniuk Petro Demianovych[1], Rubanenko Oleksandr Evgeniovich[1*] and Rubanenko Olena Oleksandrivna[2]

1 Department of Electrical Stations and Systems, Vinnytsia National Technical University, Vinnytsia, Ukraine

2 Department of Electrical Engineering Systems, Technologies and Automation in Agriculture, Vinnytsia National Agrarian University, Vinnytsia, Ukraine

*Address all correspondence to: rubanenkoae@ukr.net

References

[1] Jakushokas R, Friedman EG. Power network optimization based on link breaking methodology. IEEE Transactions on Very Large Scale Integration (VLSI) Systems. 2013;**21**(5): 983-987. DOI: 10.1109/TVLSI.2012.2201186

[2] Tirupathi R, Gulati A, Khan M, Koul R. Application of phase shifting transformer in Indian power system. International Journal of Computer and Electrical Engineering. 2012;**4**(2):242-245. DOI: 10.1109/ICGT.2012.6477970

[3] Kolcun M. Transformer use for active power flow control in the electric power system. In: Kolcun M, Hlubeň D, Beňa L, Djagarov N, Grozdev Z, editors. 2010 9th International Conference Environment and Electrical Engineering (EEEIC). 2010. pp. 246-249. DOI: 10.1109/EEEIC.2010.5489982

[4] Bocovich M. Overview of series connected flexible AC transmission systems (FACTS). In: Bocovich M, Iyer K, Terhaar RM, Mohan N, editors. North American Power Symposium (NAPS). 2013. pp. 258-263. DOI: 10.1109/NAPS.2013.6666915

[5] Bahadornejad M, Nair N-KC. Intelligent control of on-load tap changing transformer. IEEE Transactions on Smart Grid. 2014;**5**(5): 2255-2263

[6] Alekseev BA. Krupnyie silovyie transformatoryi: kontrol sostoyaniya v rabote i pri revizii. M.: Energoprogress; 2010. 88 s

[7] Rassalskiy AN, Sahno AA, Konogray SP, Guk AA. Kompleksnyiy podhod k diagnostike vyisokovoltnogo oborudovaniya podstantsiy 220–1150 kV pod rabochim napryazheniem v rezhime ekspluatatsii. ElektrotehnIka I ElektromehanIka. 2010. pp. 23-25

[8] Stohnii B, Sopel MF. Osnovy monitorynhu v elektroenerhetytsi. Pro poniattia monitorynhu. Tekhnichna Elektrodynamika. 2013;**1**:62-69

[9] Stohnii B, Kyrylenko OV, Butkevych OF, Sopel MF. Zastosuvannia zasobiv monitorynhu perekhidnykh rezhymiv v OES Ukrainy pry rozviazanni zadach dyspetcherskoho keruvannia. Vol. 23. Zb. Nauk. pr. K.: IED NANU: Pratsi Instytutu Elektrodynamiky Natsionalnoi Akademii Nauk Ukrainy; 2009. pp. 147-155

[10] Buslavets OV, Lezhniuk PD, Rubanenko OE. Evaluation and increase of load capacity of on-load tap changing transformers for improvement of their regulating possibilities. Eastern-European Journal of Enterprise Technologies. 2015;**2**(8(74)):35-41. DOI: 10.1109/icgt.2012.6477970

[11] Kylymchuk A, Lezhnyuk PB, Rubanenko OE. Reduction of additional losses of electric energy in parallel operating non-uniform electrical grids taking into account non-uniformity and sensitivity. International Journal of Energy Policy and Management.– 2015. №. ;**1**:1-5

[12] Lezhniuk P, Rubanenko OY, Nikitorovych OV. Operatyvne diahnostuvannia vysokovoltnoho obladnannia v zadachakh optymalnoho keruvannia rezhymamy elektroenerhetychnykh system. Tekhnichna Elektrodynamika. 2012;**3**: 35-36

[13] Evdokimov S, Kondrashova Y, Karandaeva O, Gallyamova M. Stationary system for monitoring technical state of power transformer. In: Procedia Engineering (2nd International Conference on Industrial Engineering (ICIE-2016)). Vol. 150. 2016. pp. 18-25. DOI: 10.1016/j.proeng.2016.07.270

[14] Mohammad B, Jabali A, Kazemi MH. Power system event ranking using a new linear parameter-varying modeling with a wide area measurement system-based approach. Energies. - MDPI. 2017;**10**(1088):1-14. DOI: 10.3390/en10081088

[15] Ghullam M, Bhutto GM, Bak CL, Ali E. Controlled operation of the islanded portion of the international council on large electric systems (CIGRE) low voltage distribution network. Energies. MDPI. 2017;**10**:1-13. DOI: 10.3390/en10071021

[16] Alekseev BA. Kontrol sostoyaniya (diagnostika) krupnyih silovyih transformatorov. In: Moskva NTs ENAS. 216 s

[17] Tenbohlen S, Vahidi F, Gebauer J. Zuverlässige ist Bewertung von Leistung Transformatoren: Materials of HS-Symposium. Universität Stuttgart; 2012. pp. 1-11

[18] Hamisu U, Hizam H, Radzi MAM. Simulation of Single-Phase Shunt Active Power Filter with Fuzzy Logic Controller for Power Quality Improvement. Universiti Putra Malaysia; 2013. pp. 353-357

[19] Sunny K, Akhtar F, Solanki S, Zbigniew L, Staszewski L. Adaptive Fault Classification Approach using Digitized Fuzzy Logic (DFL) Based on Sequence Components. Sukkur IBA University; 2018. pp. 1-7

[20] Devi S, Nair MG, Nair DR, Ilango K. Tariff Based Fuzzy Logic Controller for Active Power Sharing between Microgrid to Grid with Improved Power Quality. Amritapuri, India: Amrita School of Engineering; 2016. pp. 406-409

[21] Moudud A, Roy NK. Design of a Power System Stabilizer using Adaptive Neuro Fuzzy Logic for a Multi-Machine System having Dynamic Loads. Bangladesh: Technology University Gopalganj; 2016. pp. 402-407

[22] Priyanka R, Sinha AK. Modelling and Simulation of Grid Power Management with Modified Fuzzy Logic Based MPPT Tracking with Hybrid Power Energy System. Agartala Tripura, India: Tripura University; 2015. pp. 1-6

[23] Suna H, Huanga Y, Huang C. Fault diagnosis of power transformers using computational intelligence: A review. Energy Procedia;**14**:1226-1231. DOI: 10.1016/j.egypro.2011.12.1080

[24] Velásquez RMA, Lara JVM. Expert system for power transformer diagnosis. Engineering and Computing:434-437. DOI: 10.1109/INTERCON.2017.8079640

[25] Lin D, Xu YY, Liang Y, Li Y, Liu N, Zhang G. A risk assessment method of transformer considering the economy and reliability of power network. In: 1st International Conference on Electrical Materials and Power Equipment. pp. 594-597. DOI: 10.1109/ICEMPE.2017.7982167

Section 7

Fuzzy Logic Applications in Management

Chapter 7

The Fuzzy Logic Methodology for Evaluating the Causality of Factors in Organization Management

Nazarov Dmitry Mikhailovich

Abstract

The paper is concerned with solving the problem of factor causality using the tools of the fuzzy set theory. The paper formulates the problem of causal relations in a broad sense and analyzes the methods for its solution with an emphasis on the socioeconomic aspects. For this purpose, the system approach, comparative experiment, economic and mathematical modeling, and other general scientific methods are used. The authors suggest that the causality of factors be studied based on the theory of fuzzy binary relations using the mathematical tools of Goguen's fuzzy implication. As an example, the paper describes the effect of organizational culture indicators under the Denison's model on the key performance indicators of an organization.

Keywords: factor causality, fuzzy evaluation, causality, fuzzy binary relations, corporate culture, balanced scorecard

1. Introduction

The three well-known mathematical theories concerned with the modeling of economic systems under uncertainty are the theory of probability, the theory of possibilities, and the fuzzy set theory. Fuzzy mathematical models represent a new and promising trend in applied mathematics and are increasingly being used in various applied fields in situations involving various kinds of uncertainties where these cannot be strictly formalized by the methods of probability theory and mathematical statistics. These uncertainties can be of various types: those caused either by the inability to determine the values of parameters with mathematical precision or by the impossibility of finding their exact numerical limits.

The main idea of this tool is that any economic indicator is interpreted as an integral one and is thus defined not by an absolute number but by a certain interval (i.e., fuzzified) which corresponds to the real-life situations where only the limit values of the analyzed indicator within which it can vary are known with sufficient precision, but there is no quantitative or qualitative information about the possibilities or probabilities that its various values will be implemented within a given interval. That is, when using the mathematical apparatus of the fuzzy set theory, it is necessary to formalize one's vision of the possible values that the indicator in question can take and specify the set of its values and the degree of uncertainty that each of these values will be assumed. Once the input economic indicators are formalized, we may calculate the possibility distribution of a generalized indicator or a system of desired output characteristics by the "level principle of generalization" or "Zadeh's generalization principle." After such calculation and having built the so-called inference engine for the main economic indicators, it is necessary to defuzzify and interpret them based on the system of rules set by the researcher.

Development of such models for economic system management makes it possible to address the uncertainty in economic agents' behavior and, thus, to minimize the "human-sized" risks of decision-making.

2. Analyzing the mechanisms of management factor causality in organizations

The central goal of most scientific research is to elucidate the cause-and-effect relationships among variables or events. For millennia, the issues of "causality" have enjoyed great interest among representatives of many sciences: philosophy, psychology, economics, physics, chemistry, etc. In social and economic sciences, the "cause-and-effect relationship" issues are associated with the new term "causality" which is increasingly used. Causality (Lat. *causalis*) is a cause-and-effect relationship: a causal interdependence of events in time [26]. To establish the causal relations among the variables (synonyms "deterministic," "causal") is, probably, one of the most important problems in the scientific research. Indeed, any scientist seeks to identify a cause-and-effect relationship and implement the most effective mechanism to achieve the desired outcome. The broad application scope of the causality concept dictates the diversity of approaches to its study [3, 28].

In a broad sense, the causality theory essentially answers the simple question associated with verifying the statement that "event X generates event Y." In this case, X is called the cause or a causal factor, and Y is the consequence, response, or the resultant factor. Mathematically speaking, X is a necessary condition for Y, and Y is a sufficient condition for X. The problem of the causality theory is presented in the form of a graph below (see **Figure 1**).

Based on the studies [27, 28], the problems of causal relations among factors can be formulated as follows:

1. The problem of X directly affecting Y: does factor X actually affect factor Y directly, or is there some indirect impact that factor X exerts on factor Y through some factor Z?

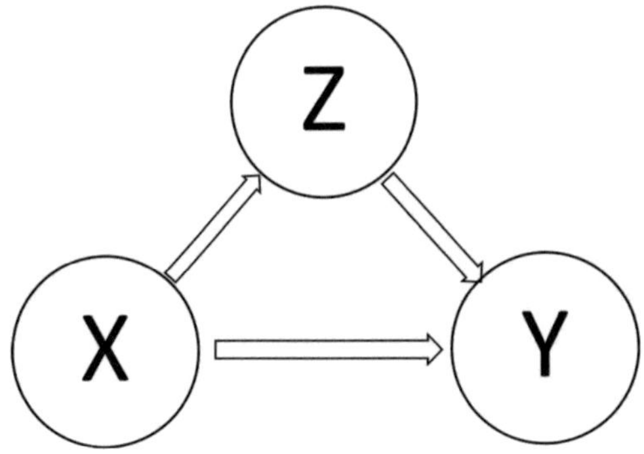

Figure 1.
Basic factor causality graph.

2. The problem of delayed or retrospective causality: did factor X actually affect the resultant factor Y, or was this effect random, and is the change in factor Y due to other reasons? Delayed causality comes into play when factor Y is measured after some time has passed since factor X, and when factor Y is measured at a given moment in time based on retrospective measurements of factor X, this is commonly referred to as retrospective causality.

3. The problem of functionality of causal relations consists of finding a solution to dichotomy: is the relation deterministic or probabilistic? In the first case, we can talk of a law, principle, etc. in the area of interest, while in the second case, it is a stable, strong, or weak relation for a certain class of objects.

The three interrelated scientific problems of causal relations can be represented in the form of a causal field. In this case, we are exploring all adjacent relationships, i.e., the set of variables {X} capable of affecting the outcome Y, the set of other resultant variables {Y} dependent on X, and the set of variables {Z}.

A causal field characterizes a set of factors that, on the one hand, provide a sufficiently complete description of the subject and make it possible to explain the obtained or predicted results based on the established interrelationships, on the other. The structure of the causal field is commonly built on the basis of substantive considerations and the experimental results, and the alleged causality is either confirmed, and hypotheses about the nature of the given and associated relationships are suggested or refuted.

The issues of causal relations in economics and management are discussed in many works by Nobel laureates who build their models on the assumption that if *Y* could exist, then *X* could have had an effect, and a situation could arise where *X* caused *Y*. That is, the authors essentially reveal the significant patterns in the causal field, test them against examples, and summarize them into economic laws. G. Akerlof argues that the simple Pareto efficient equilibrium market trading results could be radically changed if we considered buyers and sellers with a certain assumed degree of information asymmetry. Thus, G. Akerlof's model is a causal field: unless the asymmetry of market information is taken into account, the behavior of buyers and sellers will provide a less reliable description of the real market situation, and, therefore, a new theory of market behavior will need to be developed. Another prominent example in this matter is the theory of segregation. According to Schelling's theory, a causal field is a combination of at least three factors: targeted state policy, preferences of an individual market agent, and segregation. The causal relations that he established counter the commonly accepted view that segregation can only result in the targeted government policies or strong segregation preferences.

The probabilistic and statistical models involving, above all, the study of events occurring in the course of experiments are considered as the fundamental economic and mathematical models of causal relations among factors.

The first group of models implements correlation and regression analysis. Within this group of models, we should note S. Wright's structural equations and diagrams [13, 16], the Neyman-Rubin causal models [15], Pearl's functional models [14], David's dynamic models [3], and various graph models. Anyway, all these models employ different types of correlation analysis as a measure of determination and calculate a coefficient of correlation. In case of assurance that the data have a normal distribution and are of interval nature, the Pearson correlation coefficient is used; in the event of dichotomy and the use of ranks, the rank correlation coefficient or the point-biserial one is used. To identify causality, parametric and nonparametric single-factor analyses of variance are also used, and the chi-square value of the contingency coefficient in the crosstab tables is calculated and analyzed.

If an indirect effect needs to be identified, the authors, as a rule, suggest using the various modifications of the two-factor analysis of variance and the multiple regression implementation technology, where the regression line inclination is taken as a characteristic of the causal relation strength.

Structural modeling proves to be very useful in determining the significance of an indirect effect. This usually involves the comparison of two models, and their coefficients are used to estimate the indirect, direct, or ambivalent effect of factors.

The second group of models received a general name of confirmatory analysis [10, 24]. Confirmatory factor analysis, or the measurement model, relies on the assumption that relationship among several explicit variables results from the common cause of their joint variability, i.e., a factor as a latent variable. The measurement model makes it possible to test the assumption that this set of indicators is sufficient to measure the latent construct and to determine what contribution each indicator makes to its evaluation. The confirmatory factor analysis based on the method of maximum likelihood is a special case of the modeling method using linear structural equations. Unlike the exploratory factor analysis, it correlates the isolated factor structure with the one already known to or assumed by the researcher and determines the reliability of this correlation. The researcher here needs to have some idea of the test variables structure and of the causal field in general. Such idea can be defined by some theoretical principles tested in the experiment or obtained by exploratory factor analysis.

The methodology of confirmatory analysis is to a greater extent based on deductive (confirmatory) logic than on the inductive (exploratory) one. Deductive logic starts from building a structural model of directional and nondirectional relations between the given constructs with a view to its further verification for consistency with the empirical data and adjustment by means of data analysis.

The basic idea of confirmatory analysis is not only to single out a certain set of factors but also to correlate this solution with the one established previously. The starting point for this procedure is the following vector algebra equation:

$$\mathrm{C}_{kk} = L_{kf} C_{ff} L'_{fk} U_{kk}. \tag{1}$$

where C_{kk} is a covariance matrix for k variables; alternatively, a correlation matrix can be used; C_{ff} is the covariance matrix for f factors. Generally, this matrix is assumed to be diagonal, i.e., factors do not correlate with each other. L_{kf} is the factor loading matrix, and L'_{fk} is its transposed version. U_{kk} is a diagonal covariance matrix for the variable-specific factors. The presence of this parameter obviously indicates that the maximum likelihood procedure is based on the common factor model.

Several examples of how the methods described above that were used in various studies are provided below.

In her study, Yudaeva discusses the causal relation between the process of Russia's accession to the WTO and the implications thereof. The causal field is an acyclic graph whose nodes are events and arcs are information flows evaluated in a probabilistic form. Based on the expert knowledge processed by the method of randomized probabilities, a forecast is made, and the strength of relations in the acyclic graph is established on the example of the Russian electricity sector. Further, the study discusses the scenarios of possible consequences based on processed expert opinions and constructs a decision tree. Based on the aggregated expert knowledge, a method has been obtained to estimate the probability that certain alternatives will be implemented which the author interpreted as the effect of Russia's accession to the WTO on the electricity industry segment.

Since one of our key objectives is to identify and evaluate the effect of implicit factors on organizational performance, in our study we will consider the existing models and methods for evaluating the impact of such implicit factor as organizational culture [23].

It should be noted that the number of foreign and domestic studies on evaluating the impact that organizational culture has on company performance is extremely small so far and does not reflect the business needs in this respect.

The main foreign studies are the *Corporate Culture and Performance* by Kotter and Heskett [11], *Built to Last* by Collins and Porras [2], *The Balanced Scorecard* by Kaplan and Norton [20], and *Practice What You Preach: What Managers Must Do to Create a High Achievement Culture* by Maister [21]. As to research by Russian scientists, there are the studies by Solomandina, Zhuravleva, and Zhukov, who also tried to evaluate from different perspectives the effect of organizational culture on various aspects of company performance in the Russian context.

The most systematic research in this area was performed by the US scientist D. Maister, who tried to formalize the causal relationship between "organizational culture and company performance" based on the conceptual provisions of Kaplan and Norton.

Maister highlighted the following factors or elements of organizational culture: development; coaching; psychological climate in the team; high standards by which he understood the personal qualities of employees, commitment, and high performance; long-term orientation; empowerment; fair compensation; and employee satisfaction.

This set of organizational culture indicators is not accidental. From the author's point of view, it determines the financial success of an organization. The author, relying on the above works, proves this on practice by studying a fairly large number of different companies over a number of years.

Of interest is also the logic behind Maister's causal relations which is expressed as the following chain: financial indicators, product quality, and employee satisfaction (high standards, support, coaching, and empowerment) (system of fair compensation, employee commitment, and enthusiasm; coaching) (long-term orientation, interest and enthusiasm of employees). Note that some elements of the chain contain not one but two or more variables that are in turn related to each other.

Financial component of organizational performance, according to Maister, is an integral weighted convolution of the four indicators: profit per employee, revenue growth over the past 2 years, profit growth over the past 2 years, and return on sales.

By the statistical analysis of financial performance for 139 offices employing 5589 people, Maister proved that there are two factors that have the greatest influence on financial success—i.e., profit growth (0.81) and profit per employee (0.53). The remaining factors have a significantly lower impact: 0.27 and 0.24, respectively.

The relationship between financial performance and organizational culture can be analyzed using Maister's flowchart (see **Figure 2**). The respondents rate each indicator of the organizational culture on a 1–6 scale. The average score is then used for the analysis, and the relation coefficients expressed as parameter b represent an amount of change in the variable that would result from a one-unit change in another variable.

The weakness of this research lies in the use of econometric apparatus: the author assumes that connections are linear and evaluates their strength; moreover,

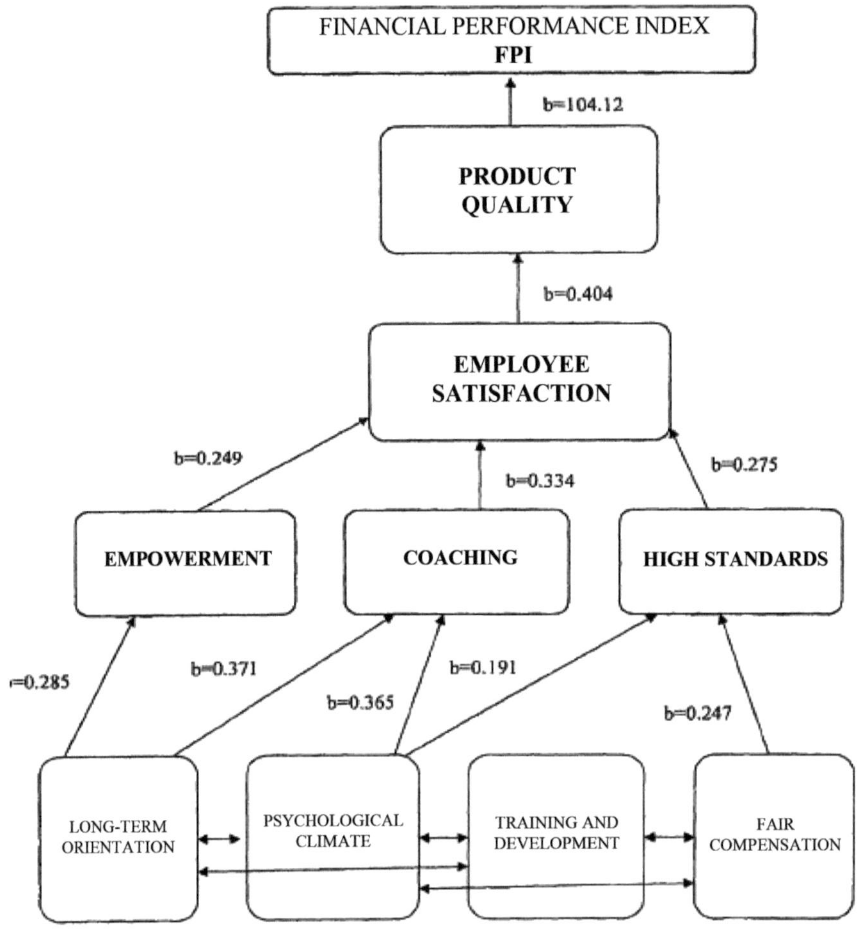

Figure 2.
Financial performance index vs. organizational culture elements [21].

the established hierarchy structures the relationships within the corporate culture factors considerably and essentially leaves out the possibility of their simultaneous independent change. In general, Maister gives a sufficiently complete and objective evaluation of the impact that the organizational culture has on the firm's financial success.

The next stage in the development of the "organizational culture vs. company performance" dual relationship was the model suggested by Professor Denison at the International Institute for Management Development in Lausanne, who, based on statistical data from more than 1000 firms, made another attempt to describe logical chains between the components of organizational culture and main performance indicators of an organization. According to this model, organizational culture is a synergistic sum of four dimensions: involvement, consistency, adaptability, and mission. Each dimension is broken down further in accordance with the research; in particular, the mission determines the strategy, goals, and objectives, as well as the company's vision; consistency involves coordination/integration, agreement, and core values; involvement comprises team orientation, capability development, and empowerment; and adaptability involves creating change, customer focus, and organizational learning [18].

In the Denison's model, the set of financial performance indicators of an organization has also been significantly expanded: in addition to financial indicators, he suggests using a wide range of key measurable and qualitative performance indicators such as assets and investments, sales and product quality, employee satisfaction and innovation, creativity and customer focus, sales growth, and market share gain. This approach provides a more complete reflection of the relationship between organizational culture and performance in the broad sense. That is, Denison's model combines the ideas of Kravetz, Thompson, and Maister and contains its own new features, being a more accurate tool to determine what impact the organizational culture has on company performance. In essence, Denison suggested his own original causal field of factors that link organizational performance with culture.

Denison found that mission and consistency have a greater impact on financial indicators, such as ROA (return on assets), ROI (return on investment), and ROS (return on sales). The value of the mission and consistency indices of three to four usually indicates a high return on investment, assets, and sales, as well as the operational strength of an organization.

Consistency and involvement (internal focus) affect quality, employee satisfaction, and return on investment. Similarly, the value of these indices from 3 to 4 indicates high product quality, smaller percentage of scrap and rework, proper resource allocation, and a higher level of employee satisfaction.

Involvement and adaptability influence product development and innovation. When these parameters have an index of three to four, it means a high level of innovation in manufacturing and service, creativity, and a quick response to the changing desires and needs of both clients and their own wage workers.

Adaptability and mission (external focus) influence revenues, sales growth, and market share. When the values of these parameters range from 3 to 4, the organization is likely to see a steady sales growth and market share gain.

Denison's method for index calculation is based on statistical estimates that get recalculated when a new object of research is added to the knowledge base and refines the impact standard derived from changes in the data set.

In the Russian research of recent years, we can note the study by Zhuravleva who developed her own impact evaluation model (causal field) with regard to the specifics of the Russian business activity (**Figure 3**).

As know-how of her model, the author suggests using the structural elements of "effective leadership," "horizontal management structure," and "loyalty and creativity at work" influencing such performance indices as the product quality, sales growth, employee turnover, labor productivity, and the number of labor misconducts. From the author's point of view, organizational culture can also be evaluated using the following indices: creativity factor, innovation level, coefficient of satisfaction with the organization, rate of knowledge and skill implementation, worker qualification factor, professional competence factor, and responsibility factor [22].

Given the specifics of the Russian economy and the socio-morphic nature of the organizational culture (OC) phenomenon, Russian scientists note its influence on the product quality, sales growth, employee turnover, labor productivity, number of labor misconducts while evaluating OC through the creativity factor, the level of innovation, the coefficient of satisfaction with the organization, and the factors of knowledge and skill implementation, employee qualification, professional competence, and responsibility.

In her study, Pervakova [25] builds the following flowchart of influence that the organization culture has on business performance and labor productivity (**Figure 4**).

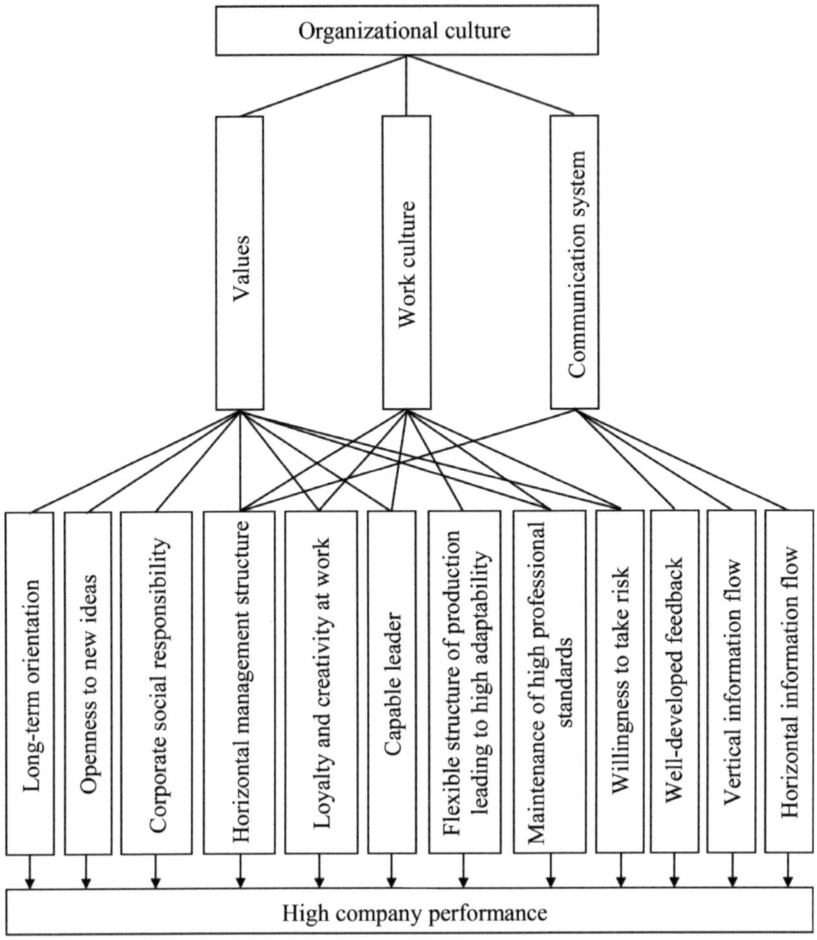

Figure 3.
Structural elements of the organizational culture which determine the company's performance [19].

The literature review once again proved the scale of the problem being addressed. The research revealed a sufficiently large number of parameters and factors of influence and determined their principles and mechanisms. The authors of all models, both domestic and foreign, only determine the dichotomous effect of organizational culture on the key performance indicators describing the qualitative influence (a typical example is with an effective organizational culture, the turnover rate is low) while not trying to formalize it in order to answer the question of both how and to what degree (weakly, neutrally, strongly) the level of OC affects the parameters of interest. Note also that even when quantitative estimates are used in the studies, they are based on econometric relationships and, therefore, assume the existence of a serious database of accumulated results. And finally, one of the most important observations: a number of authors believe the organizational culture components to have a direct impact on financial and economic performance, while others are confident that this impact is indirect. We adhere to the latter point of view, and our evaluation model will be based on this very assumption.

The experience of foreign researchers and practitioners who studied how the organizational culture affects business performance suggests that:

1. The organizational culture has a *direct* prevailing effect on *employee satisfaction*, *job involvement*, and *ethics of customer communication*. These factors in turn

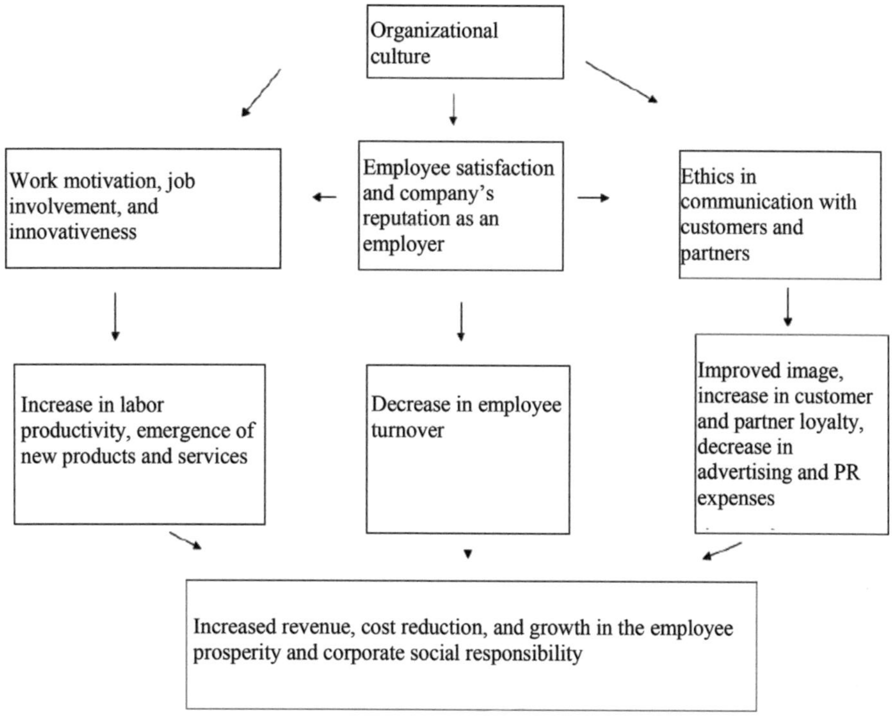

Figure 4.
Flowchart of the OC impact on business performance.

affect the financial and market performance of companies which they interpret as customer satisfaction and loyalty and labor productivity and profitability [4–7, 9].

2. The types of organizational culture also affect the financial and market performance of companies, for example, in the Cooke and Rousseau model [26], the cooperation and competition types are most favorable to the financial and market performance.

3. Relying on the OSP model of organizational culture, Sheridan and Chatman discovered that some of its components such as "respect for people" and "team orientation" have a much greater effect on job involvement, employee satisfaction, and decrease in staff turnover, than the others [1].

4. The most advanced in this respect is the Denison model. It was his practical experience that allowed him to identify those components of organizational culture that affect performance indicators of a generalized business unit.

The key difference of our methodology lies in formalizing a factor as a linguistic variable and in using standard fuzzification and defuzzification techniques to come to a conclusion based on fuzzy logic procedures.

The search for and interpretation of the information related to building a causal field, to the identification of factors, and to the meaningful evaluation of obtained result remain beyond the scope of the literature reviewed above. The model of the causal field being suggested by the author is an acyclic graph, and the strength of relations is calculated from processing the expert judgments by traditional graph-matrix techniques.

3. Fuzzy model to evaluate the causality of organization's performance factors

Linguistic variable [12] as a special tool in the fuzzy set theory allows us to formalize the verbal description of a balanced scorecard and its structural properties but disregards the strength of relations between indicators and factors within it. Therefore, in our opinion an adequate model to evaluate the impact of implicit factors on economic processes should be developed based on a combination of the fuzzy set theory concepts and objectification of the expert judgments.

The following model allows us to find the degree to which an implicit factor influences the measurable ones, i.e., those whose values can be obtained quantitatively by introducing an indirect factor into the model.

To develop a model, one needs at least to explore the three sub-models that make up the economic system. For each sub-model, we introduced the following designations: A, implicit factors; B, indirect indicators; and C, quantitatively measurable indicators.

The general plan for model development consists of two steps:

First step: to develop sub-models

Second step: to combine sub-models into a general model, to analyze it, and to address the problem in question

Sequence of operations at the first step:

- Initial determination of a set of numerical indicators for each sub-model
- Lists of sets of numerical indicators

Sequence of operations at the second step:

- Evaluation of the mutual influence among indicators in pairs (A, B), (A, C), and (B, C)
- Finding the indirect effects of model A indicators on model C indicators
- Interpretation of the obtained results

3.1 Sub-model development

The set of selected implicit factors will be considered as a carrier set of sub-model A, that of indirect factors as sub-model B and that of the measurable factors as sub-model C.

Sub-models A, B, and C can be represented by the sets of indicators:

$$\begin{aligned} A &= \{a_1, a_2, \dots, a_n\}, \\ B &= \{b_1, b_2, \dots, b_m\}, \\ C &= \{c_1, c_2, \dots, c_k\}. \end{aligned} \tag{2}$$

3.2 Combination of sub-models into a general model and its analysis

To identify the latent, B-mediated effects of A indicators on C indicators, one can use a combination of a hierarchy analysis method and the fuzzy relation theory.

In this case, we are interested in fuzzy binary relations:

$$
\begin{aligned}
a\rho_1 b &: a \text{ affects } b,\ (a,b) \in A \times B,\\
b\rho_2 c &: b \text{ affects } c,\ (b,c) \in B \times C,\\
a\rho_3 c &: a \text{ affects } c,\ (a,c) \in A \times C
\end{aligned} \tag{3}
$$

The fuzzy relation theory can be applied to identify and evaluate the implicit effects. Relations are given by the matrices J_{AB}, J_{AC}, and J_{BC}, whose elements are the values of membership for the corresponding pair of elements in a binary relation. The definition of membership functions is known to be the most difficult part of the fuzzy set theory. This is where the hierarchy analysis method can be of help.

Assume, for example, that matrix J_{AB} is given as

$$
J_{AB} = \begin{pmatrix} s_{11} & s_{12} & \dots & s_{1m} \\ s_{21} & s_{22} & \dots & s_{2m} \\ \dots & \dots & \dots & \dots \\ s_{n1} & s_{n2} & \dots & s_{nm} \end{pmatrix}, \tag{4}
$$

where $s_{ij}\left(0 \le s_{ij} \le 1; \quad i = 1, 2, \dots, n; \ j = 1, 2, \dots, m\right)$ is the strength of effect that indicator a_i has on indicator b_j.

The s_{ij} values are usually determined by experts. The analytic hierarchy process (AHP) can be used here for the purposes of consistency and clarification and to increase the validity of expert judgments related to s_{ij} values.

The diagram of hierarchies in this case has the following form (**Figure 5**).

Calculated by the standard procedure, the normalized estimates of the priorities vector for each b_j should be written in jth column of the J_{AB} matrix: $\left(s_{1j}, s_{2j}, \dots, s_{nj}\right)^{\mathrm{T}}$.

However, remember that the resulting J_{AB} matrix would only reflect the expertly established effect of indicator a_i on indicator b_j if all the normalized estimates of each priority vector meet the chosen consistency measure. Otherwise, either the models A and B themselves or the expert judgments will need to be revised.

The J_{AC} and J_{BC} matrices are composed in a similar way:

$$
J_{AC} = \begin{pmatrix} z_{11} & z_{12} & \dots & z_{1k} \\ z_{21} & z_{22} & \dots & z_{2k} \\ \dots & \dots & \dots & \dots \\ z_{n1} & z_{n2} & \dots & z_{nk} \end{pmatrix}, J_{BC} = \begin{pmatrix} u_{11} & u_{12} & \dots & u_{1k} \\ u_{21} & u_{22} & \dots & u_{2k} \\ \dots & \dots & \dots & \dots \\ u_{m1} & u_{m2} & \dots & u_{mk} \end{pmatrix}, \tag{5}
$$

where $z_{ij}\ \left(0 \le z_{ij} \le 1; \quad i = 1, 2, \dots, n; \quad j = 1, 2, \dots, k\right)$ is the strength of effect that indicator a_i has on indicator c_j and
$u_{ij}\ \left(0 \le u_{ij} \le 1; \quad i = 1, 2, \dots, m; \quad j = 1, 2, \dots, k\right)$ is the strength of effect that indicator b_i has on indicator c_j.

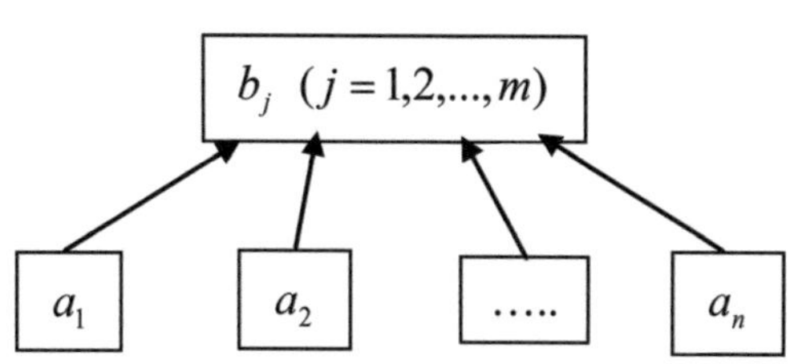

Figure 5.
The diagram of hierarchies in AHP.

The latent, B-mediated effects of sub-model A indicators on those of sub-model C can be established and evaluated as follows (**Figure 6**).

The strength of direct effect that a_i has on c_i is determined by matrix element z_{i1}. Similarly, the strength of direct effect that a_i has on $c_2, \ldots, c_k$ is set by the numbers $z_{i2}, \ldots, z_{ik}$ in this matrix. In addition to direct impact, the indicator a_i affects $c_1, \ldots, c_k$ through the mediating element b_i, a sub-model B indicator. The strength of b_i-mediated *indirect* impact of a_i on $c_1, \ldots, c_k$ is set to $z^*_{i1}, z^*_{i2}, \ldots, z^*_{ik}$ values that represent the minimums of s_{ij} and correspond to $u_{j1}, u_{j2}, \ldots, u_{jk} : z^*_{i1} = \min(s_{ij}, u_{j1}), z^*_{i2} = \min(s_{ij}, u_{j2}), \ldots, z^*_{ik} = \min(s_{ij}, u_{jk})$. The a_i element can affect each of the $c_1, \ldots, c_k$ elements not only through the "mediator" b_i but also through any element of sub-model B (**Figure** 7).

The cumulative mediated effect of element a_i on c_j is set equal to the maximum effect mediated through all the elements of sub-model B:

$$z^*_{ij} = \max(\min(s_{i1}, u_{1j}), \min(s_{i2}, u_{2j}), \ldots, \min(s_{im}, u_{mj})), \tag{6}$$

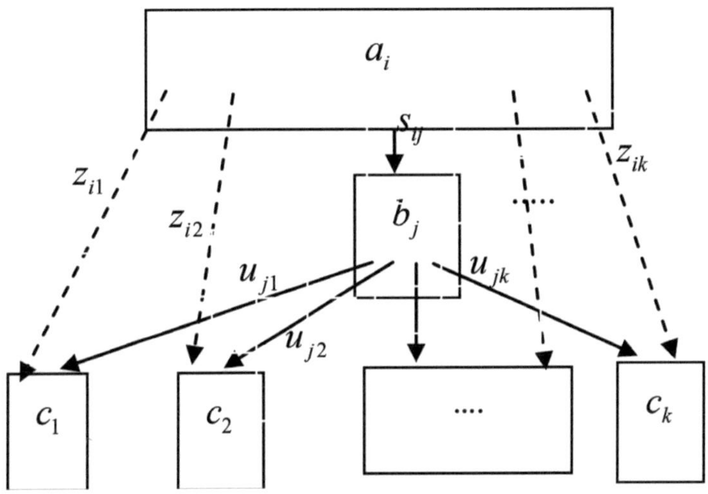

Figure 6.
Direct and b_j*-mediated impact of* a_i *on sub-model* C *elements.*

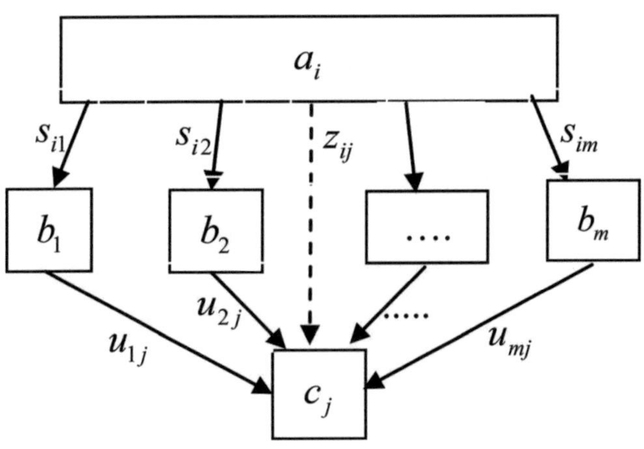

Figure 7.
Direct and sub-model B *element-mediated effect of* a_i *on* c_j.

Considering the operation "min" as multiplication and "max" as an addition, it appears that all the *B*-mediated effects of *A* on *C* are defined in the product of matrices J_{AB} and J_{BC}:

$$J_{AC}^{*} = J_{AB} \cdot J_{BC} = \begin{pmatrix} z_{11}^{*} & z_{12}^{*} & \dots & z_{1k}^{*} \\ z_{21}^{*} & z_{22}^{*} & \dots & z_{2k}^{*} \\ \dots & \dots & \dots & \dots \\ z_{n1}^{*} & z_{n2}^{*} & \dots & z_{nk}^{*} \end{pmatrix}, \tag{7}$$

where z_{ij}^{*} is defined by Eq. (1).

All the values of s_{ij}, u_{ij}, z_{ij} are expertly determined.

If the strength of direct effect of *A* on *C* expertly determined by the analytic hierarchy process exceeds the indirect one, then there is no point in accounting for it. If the inequality $z_{ij}^{*} - z_{ij} > 0$ holds, then an indirect (and not recognized by experts) effect of the *i*th implicit factor on the *j*th resulting index is found. Moreover, the difference $z_{ij}^{*} - z_{ij}$ can be considered an estimated strength of such effect.

The developed model makes it possible to find and evaluate the strength of the indirect effect that implicit factors have on the system's key measurable indicators. It combines two mathematical techniques, i.e., the analytic hierarchy process and the theory of fuzzy binary relations. Each of these techniques is quite widely used, but we have not found their combination in the available literature. The quantitative and qualitative conclusions derived from this model are easy to interpret and verify in practice.

The developed model makes it possible to expertly find and evaluate the strength of the indirect effect of implicit factors on the system's key measurable indicators. In the thesis, we suggest that the estimated effects be obtained using Goguen's fuzzy logic inference, as it is the one that satisfies the logic of defining the mutual effects among indicators within the sub-models A, B, and C.

Step 1. At this step, the effect matrices $J_{AB}; J_{BC}$ are structurally and quantitatively defined according to the rules R1 and R2:

$$R1: J_{AB} = \{x_{ij}\} = \left(min\left\{1, \frac{b_j}{a_i}\right\} \right), \text{ where } i = 1..n, j = 1..m. \tag{8}$$

$$R2: J_{BC} = \{y_{jk}\} = \left(min\left\{1, \frac{c_k}{b_j}\right\} \right), \text{ where } j = 1..m,\ k = 1..k. \tag{9}$$

The resulting matrix that estimates the effects between sub-models *A* and *C* is found by the rule of minimax matrix products:

$$J^{*} = J_{AB} \cdot J_{BC} \tag{10}$$

Step 2. At this step, the indicators of sub-models A* and C* are recorded after the implicit factor is changed, while the effect matrices defined by Eqs. (3), (4), and (5) remain the same (the strength of effects does not change).

Step 3. Calculation of the quantifiable indicators of sub-model $C_{estimate}$:

$$C_{estimate} = J^{*} \cdot A^{*}, \tag{11}$$

where A^{*} is a set of numerical values of the implicit factor measured and recorded after the change.

Assume that C^* is the set of numerical values for the measurable indicators of sub-model C after the implicit factor is changed. The paper suggests that the $C_{estimate}$, C^* indicator sets be defuzzified by Mamdani algorithm into $dC_{estimate}$, dC^* and the relative error be found for the defuzzified difference values. The $\frac{dC^* - dC_{estimate}}{dC_{estimate}}$ indicator will be the desired estimated effectiveness of the proposed impact evaluation. Relations are given by matrices J_{AB}, J_{BC}, J_{AC} whose elements are the values of membership of the corresponding pair of elements in a binary relation.

4. Implementation of the fuzzy model to evaluate the causality of organization's performance factors

Using the reflexive selection procedure, a causal field was built, and indicators were divided into three groups according to the model. The expert distribution is presented in the form of **Table 1**.

We have found that KK_1 (adaptability) has little direct effect on OP_1, OP_2, and OP_3 but a strong OP_1-mediated one on PP_4. Other indirect effects were identified in a similar way.

As for the mathematical model, we introduced the following notation: fuzzy set $A = \{KK_1, KK_2, KK_3, KK_4\}$ describes the indicators of the IT company's organizational culture, fuzzy set $B = \{PP_1, PP_2, PP_3, PP_4\}$ describes indirect indicators, and fuzzy set $C = \{OP_1, OP_2, OP_3\}$ comprises the IT company's key performance indicators.

Note that the indicators chosen for the purposes of model implementation have different units of measurement. Therefore, they need to be modified by being presented as a fuzzy set. In our model, after looking into different normalizing methods to represent each sub-model's indicators as a fuzzy set, we found little to no variation in results as a function of the data normalizing method, and, therefore, each sub-model was assigned the membership function obtained by normalizing the intra-sub-model indicator values by means of dividing them by the maximum indicator. The obtained estimates are interpreted as a degree to which the indicators influence each other within the set. At the same time, we should understand that all these indicators (within each sub-model) must be measured in the same units (rubles, percent, fractions, etc.).

Therefore, in the course of the experiment, we selected performance indicators of the target organizations, revised their structures using reflexive selection model, and built their causal field with regard to the structure defined above.

A fuzzy evaluation of the effect that an implicit factor has on organization's key performance indicators will be obtained using the fuzzy logic rules, algorithms, and

Organizational culture indicators (implicit factor) (sub-model *A*)	Indirect (intermediary) indicators (sub-model *B*)	Key performance indicators of an IT company (sub-model *C*)
KK_1: adaptability KK_2: mission KK_3: cooperation KK_4: involvement	PP_1: percentage of innovative solutions in services and sales PP_2: percentage of projects performed on time PP_3: percentage of proceeds from each customer PP_4: percentage of innovations per employee	OP_1: net profit OP_2: sales of products and services OP_3: rate of return

Table 1.
System of IT company performance indicators divided into three groups.

procedures, but J. Goguen's[1] fuzzy implication will be taken as a basis since it satisfies the logic of relations among our indicators within the developed causal model.

The fuzzy inference rules by which we will evaluate the strength of relations among indicators as elements of fuzzy sets [17] that we defined in **Table 1** can be written as follows:

R1: If $KK_1 = a_1$ and $KK_2 = a_2$ and $KK_3 = a_3$ and $KK_4 = a_4$, then $PP_1 = b_1$ and $PP_2 = b_2$ and $PP_3 = b_3$ and $PP_4 = b_4$.

R2: If $PP_1 = b_1$ and $PP_2 = b_2$ and $PP_3 = b_3$ and $PP_4 = b_4$, then $OP_1 = c_1$ and $OP_2 = c_2$ and $OP_3 = c_3$.

According to the theory of fuzzy binary relations, each rule can be represented in the form of matrix:

$$\mathrm{R1}: J_{AB} = \{x_{ij}\} = \left(min\left\{1, \frac{b_j}{a_i}\right\}\right) \text{where } i = 1..4, j = 1..4. \quad (12)$$

$$\mathrm{R2}: J_{BC} = \{y_{jk}\} = \left(min\left\{1, \frac{c_k}{b_j}\right\}\right) \text{where } j = 1..4, k = 1..3. \quad (13)$$

The final impact evaluation matrix can be found according to the minimax matrix product principle:

$$J^* = J_{AB} \cdot J_{BC} \quad (14)$$

This matrix shows the extent to which implicit factor indicators influence key performance indicators of an organization. Using this matrix, we can estimate the cost of improving the implicit factor (in our case, OC) based on changes in the key performance indicators of the organization.

Note that all the indicators (a_i, b_j, c_k) necessary to evaluate the strength of relation are based on the current state of business in the respective organizations from time to time.

Using the web service named "Implicit Factors Impact Evaluation" (bi.usue.ru), we will present the results of this technique being applied to all three organizations under study.

Application of the mechanism through which the implicit factors affect the key performance indicators of an organization is described through the example of OOO nanoinform. **Table 2** shows the performance indicators of OOO nanoinform for October-November 2014. Based on these indicators, the membership functions were constructed.

To obtain the values of the fuzzy set membership functions (**Tables 4** and **5**) that characterize OOO nanoinform's performance indicators, we divided each indicator in each indicator group by the maximum value for this group and obtained the values characterizing each indicator's membership degree (**Table 3**). Interpretation of the obtained results is simple and clear—it is the degree to which the indicators within a group affect each other, which meets the purpose of our model.

For the final model value for a fuzzy set *C* that characterizes the key performance indicators, we made a Mamdani fuzzy inference. It characterizes the

---

[1] J. Goguen's fuzzy implication or simply a fuzzy proposition implication in the form of ("if, then") is a binary logical operation resulting in a fuzzy proposition, the truth of which can take on the value defined by the formula:

$T(X \to Y) = \min\{1, T(Y)/T(X)\}$ where $T(X) > 0$.

Sub-model	Indicator	Value
A	KK_1, score	4.25
	KK_2, score	3.35
	KK_3, score	4.56
	KK_4, score	4.40
B	PP_1, %	20
	PP_2, %	15
	PP_3, %	17
	PP_4, %	12
C	OP_1, RUR	33,200
	OP_2, RUR	102,700
	OP_3, RUR	5100 (13%)

Table 2.
Performance indicators of OOO nanoinform for October-November 2014.

Sub-model	Indicator	Membership function value
A	KK_1	0.93
	KK_2	0.73
	KK_3	1
	KK_4	0.96
B	PP_1	1
	PP_2	0.75
	PP_3	0.85
	PP_4	0.6
C	OP_1	0.3
	OP_2	1
	OP_3	0.05

Table 3.
Membership functions based on performance indicators of OOO nanoinform.

A×B	1	0.75	0.85	0.6
0.93	1	0.81	0.91	0.65
0.73	1	1	1	0.82
1	1	0.75	0.85	0.6
0.96	1	0.78	0.89	0.63

Table 4.
Goguen's fuzzy logic rules establishing the fuzzy binary correspondences among indicators of an implicit factor (organizational culture) and indirect indicators of OOO nanoinform for October-November 2014, obtained using model formulas.

composite indicator obtained from the estimate indicators—the main ones for evaluating the company's performance. The rule is if $OP_1 = c_1$ and $OP_2 = c_2$ and $OP_3 = c_3$, then the composite indicator = c^*.

Such representation would be very convenient, as it would allow us to calculate the effect resulting from changes in implicit indicators.

After taking a number of measures aimed at improving the indicators, their new values were obtained (**Table 7**).

To summarize the above, we pooled the final indicators according to the model and the actual indicators (**Table 8**).

In general, the data presented in **Table 8** indicate the reliability of selected model, since the error was only 3%. To give a more accurate estimate of the implicit parameters' quantitative impact on the organization's key performance indicators is hardly possible, as their impact is partial and would be hard to formalize and, most importantly, to separate from other impacts. To calculate even an approximate impact, we had to "fuzzify" the intermediate indicators twice in order to prevent reevaluation of the effect that the organizational culture factors have on indicators of interest.

B×C	1	0.3	0.05
1	1	0.3	0.05
0.75	1	0.4	0.07
0.85	1	0.35	0.06
0.6	1	0.5	0.08

Table 5.
Goguen's fuzzy logic rules establishing the fuzzy binary correspondences among indirect indicators and key performance indicators of OOO nanoinform for October-November 2014, obtained using model formulas.

A×C	1	0.3	0.05
0.93	1	0.5	0.08
0.73	1	0.5	0.08
1	1	0.5	0.08
0.96	1	0.5	0.08

Table 6.
Goguen's fuzzy logic rules establishing the fuzzy binary correspondences among indicators of the implicit factor (organizational culture) and key performance indicators of OOO nanoinform for October-November 2014, obtained using model formulas.

Indicator	Indicator value before taking measures aimed at improving the indicators of the implicit factor (organizational culture), score	Membership function value before taking measures aimed at improving the indicators of the implicit factor (organizational culture)	Indicator value after taking measures aimed at improving the indicators of the implicit factor (organizational culture), score	Membership function value after taking measures aimed at improving the indicators of the implicit factor (organizational culture)
KK_1	4.25	0.93	4.29	0.96
KK_2	3.35	0.73	3.36	0.76
KK_3	4.56	1	4.45	1
KK_4	4.4	0.96	4.44	1

Table 7.
Values of the organizational culture indicators for OOO nanoinform and of the membership functions before and after taking measures aimed at improving the organizational culture indicators.

The similar studies were conducted for two other companies.

The results of research for OOO Invest Water Technology, with the first study and measurements performed in May 2014 and the final ones in December 2014, are presented in **Tables 9–14**.

Indicator designation	Indicator value before taking measures aimed at improving the indicators of the implicit factor (organizational culture), RUR	Indicator value after taking measures aimed at improving the indicators of the implicit factor (organizational culture) under the proposed model, RUR	Actual indicator value, RUR
OP_1	33,200	51,300	48,334
OP_2	102,700	102,700	112,000
OP_3	13%	18%	18%
Composite indicator	80,926	85,494	88,303
Model error, %	—	3	—

Table 8.
Comparison of the actual and model-based values for the OOO nanoinform key performance indicators.

Sub-model	Indicator	Value
A	KK_1, score	0.76
	KK_2, score	0.78
	KK_3, score	0.76
	KK_4, score	0.75
B	PP_1, %	14
	PP_2, %	12
	PP_3, %	10
	PP_4, %	9
C	OP_1, RUR	3,230,000
	OP_2, RUR	6,458,000
	OP_3, RUR	712,300 (6%)

Table 9.
Performance indicators of OOO Invest Water Technology for March 2014.

A×B	1	0.86	0.71	0.64
0.97	1	0.89	0.73	0.66
1	1	0.86	0.71	0.64
0.97	1	0.89	0.73	0.66
0.96	1	0.9	0.74	0.67

Table 10.
Goguen's fuzzy logic rules establishing the fuzzy binary correspondences among the indicators of an implicit factor (organizational culture) and indirect indicators of OOO Invest Water Technology for March 2014, obtained using model formulas.

B×C	0.5	1	0.11
1	0.5	1	0.11
0.86	0.58	1	0.13
0.71	0.7	1	0.15
0.64	0.78	1	0.17

Table 11.
Goguen's fuzzy logic rules establishing the fuzzy binary correspondences among indirect indicators and key performance indicators of OOO Invest Water Technology for March 2014, obtained using model formulas.

A×C	0.5	1	0.11
0.97	0.7	1	0.17
1	0.7	1	0.17
0.97	0.7	1	0.17
0.96	0.7	1	0.17

Table 12.
Goguen's fuzzy logic rules establishing the fuzzy binary correspondences among indicators of the implicit factor (organizational culture) and key performance indicators of OOO Invest Water Technology for March 2014, obtained using model formulas.

Indicator	Indicator value before taking measures aimed at improving the indicators of the implicit factor (organizational culture), score	Membership function value before taking measures aimed at improving the indicators of the implicit factor (organizational culture)	Indicator value after taking measures aimed at improving the indicators of the implicit factor (organizational culture), score	Membership function value after taking measures aimed at improving the indicators of the implicit factor (organizational culture)
KK_1	0.76	0.97	0.81	1
KK_2	0.78	1	0.81	1
KK_3	0.76	0.97	0.63	0.78
KK_4	0.75	0.96	0.78	0.96

Table 13.
Values of the organizational culture indicators for OOO Invest Water Technology and of the membership functions before and after taking measures aimed at improving the organizational culture indicators.

Indicator designation	Indicator value before taking measures aimed at improving the indicators of the implicit factor (organizational culture), RUR	Indicator value after taking measures aimed at improving the indicators of the implicit factor (organizational culture) under the proposed model, RUR	Actual indicator value, RUR
OP_1	3,230,000	4,522,000	3,686,345
OP_2	6,458,000	6,458,000	6,814,123
OP_3	6%	15%	686,558 (14%)
Composite indicator	5,061,973	5,490,329	5,345,735.40
Model error, %	—	3	—

Table 14.
Comparison of the actual and model-based values for the OOO Invest Water Technology key performance indicators.

The results of research for the Regional Office of OOO SAP SNG in Yekaterinburg, with the first study and measurements performed in May 2014 and the final ones in December 2014, are presented in **Tables 15–20**.

Sub-model	Indicator	Value
A	KK_1, score	0.78
	KK_2, score	0.78
	KK_3, score	0.76
	KK_4, score	0.75
B	PP_1, %	15
	PP_2, %	20
	PP_3, %	15
	PP_4, %	14
C	OP_1, RUR	4,530,000
	OP_2, RUR	7,378,000
	OP_3, RUR	933,500 (10%)

Table 15.
Performance indicators of the Regional Office of OOO SAP SNG in Yekaterinburg for May 2014.

A×B	0.75	1	0.75	0.7
1	0.75	1	0.75	0.7
1	0.75	1	0.75	0.7
0.97	0.77	1	0.77	0.72
0.96	0.78	1	0.78	0.73

Table 16.
Goguen's fuzzy logic rules establishing the fuzzy binary correspondences among the indicators of an implicit factor (organizational culture) and indirect indicators of the Regional Office of OOO SAP SNG in Yekaterinburg for May 2014, obtained using model formulas.

B×C	0.61	1	0.13
0.75	0.81	1	0.17
1	0.61	1	0.13
0.75	0.81	1	0.17
0.7	0.87	1	0.19

Table 17.
Goguen's fuzzy logic rules establishing the fuzzy binary correspondences among indirect indicators and key performance indicators of the Regional Office of OOO SAP SNG in Yekaterinburg for May 2014, obtained using model formulas.

Thus, the experimental studies demonstrated that our model satisfies actual practice requirements and provides the composite value with an error not exceeding 3%. Using simulation technology, we established that in 93% of cases, the model error does not exceed 3%.

A×C	0.61	1	0.13
1	0.75	1	0.19
1	0.75	1	0.19
0.97	0.77	1	0.19
0.96	0.78	1	0.19

Table 18.
Goguen's fuzzy logic rules establishing the fuzzy binary correspondences among indicators of the implicit factor (organizational culture) and key performance indicators of the Regional Office of OOO SAP SNG in Yekaterinburg for May 2014, obtained using model formulas.

Indicator	Indicator value before taking measures aimed at improving the indicators of the implicit factor (organizational culture), score	Membership function value before taking measures aimed at improving the indicators of the implicit factor (organizational culture)	Indicator value after taking measures aimed at improving the indicators of the implicit factor (organizational culture), score	Membership function value after taking measures aimed at improving the indicators of the implicit factor (organizational culture)
KK_1	0.78	1	0.80	0.99
KK_2	0.78	1	0.81	1
KK_3	0.76	0.97	0.70	0.86
KK_4	0.75	0.96	0.80	0.99

Table 19.
Values of the organizational culture indicators for the Regional Office of OOO SAP SNG in Yekaterinburg and of the membership functions before and after taking measures aimed at improving the organizational culture indicators.

Indicator designation	Indicator value before taking measures aimed at improving the indicators of the implicit factor (organizational culture), RUR	Indicator value after taking measures aimed at improving the indicators of the implicit factor (organizational culture) under the proposed model, RUR	Actual indicator value, RUR
OP_1	4,530,000	5,546,428	4,986,345
OP_2	7,378,000	7,378,000	7,734,123
OP_3	10%	16%	16%
Composite indicator	5,898,078	6,292,497	6,209,014.86
Model error, %	—	1	—

Table 20.
Comparison of the actual and model-based values for the Regional Office of OOO SAP SNG in Yekaterinburg key performance indicators.

5. Methods of interpreting and making managerial decisions to improve organizational performance with regard to factor causality

Business performance management (BPM) is a closed process consisting of four interrelated steps (strategy development, planning, monitoring and analysis, actions and adjustment) that transform business strategy into actions. The architecture of performance management system consists of a business component and a technical one. The common framework that binds these two components together is the measurements that define the leading, lagging, and diagnostic business performance indicators serving as performance monitoring and organization managing tools. At the present stage, the methodology of business performance management (BPM) and balanced scorecard of an organization recommends using lead indicators as they provide a wider overview of the future, expected performance, and allow managing people, processes, and technologies with lower risk. Performance management architecture is typically implemented as a panel of indicators within the balanced scorecard in the form of a multilayer application based on the business analysis and data integration infrastructure that allows the organizations to measure and monitor performance indicators more effectively and to control them.

Our proposed technique for making and interpreting managerial decisions is based on business analysis of the organization's processes, i.e., the tools and technologies necessary to transform data into information and information into knowledge and plans that ensure effective business conduct and use the following system of principles that we synthesized based on domestic and foreign experience in [8].

The overlap principle is implemented by introducing a new indicator into the model, namely, the fuzziness index that reflects the inconsistency in respondents' opinions. The use of this fuzziness index to process the respondents' answers gives us at least two advantages: the first one is due to the fact that this index is sensitive to the spread of their opinions and not sensitive to the number of respondents. This allows us to conduct these experiments even in small companies and to obtain adequate results.

The principle of openness to text is implemented in the interpretation model itself which is based on the fuzzy logic apparatus and relevant algorithms but has the field- and time-proven Denison questionnaire as its core part.

There is also the principle of effective history at work, the meaning of which lies in semantic essence of the model obtained and improved as a result of an in-depth scientific research in the global and domestic economics and mathematics.

Drawing the critics' attention to the obtained unconventional results, it is worth noting that we developed our mathematical model for calculating and interpreting data using the principle of common lexicon that allowed us to frame our own vision of results based on the available and time-proven lexicon, not contradictory in a broad sense to the conventional vision obtained under Denison's model.

The principle of lived experience and a similar principle of hermeneutic circle allow us to get completely new interpretations for the obtained data from the existing context by going "from particulars to generals" and vice versa, by going from data to information and knowledge and thus increasing the value of the obtained information due to similarity of model lexicons.

The principle of data intellectualization involves not only interpretation but also development of the specialized intelligent algorithms that can optimally transform data into information and knowledge suitable for making effective decisions.

The principles listed above are powerful tools for turning companies into self-learning organizations, where decisions that ensure progress toward strategic goals are made on the basis of facts generated using the business analysis procedures, namely, *from data to information* (turns feedstock (data) into various information products collected and aggregated as data banks), *from information to knowledge* (analytical tools identify trends, patterns, and deviations and turn information into a new product, i.e., knowledge), *from knowledge to rules* (rules that form new management institutions are formulated on the basis of laws, models, and schemes discovered by analytical tools), and *from rules to actions* (plans are developed that allow the rules to be implemented as actions of a business user).

By using the formulated principles to interpret the results obtained in the course of work, we can formulate the rules for their application to making managerial decisions and essentially obtain new formal managerial institutions for a given type of organizations.

The results of the study demonstrated that one of conditions for effective performance in the field of information technology is the existence of a mature implicit factor such as organizational culture that has a positive effect on the organization's performance. Such a strong positive relationship between implicit factor and business processes is primarily due to the fact that all indices under the Denison model are of either higher or high level achieved by a clear understanding by the staff in all the studied organizations of what for and why "they are here and now" (see **Figures 8** and **9**).

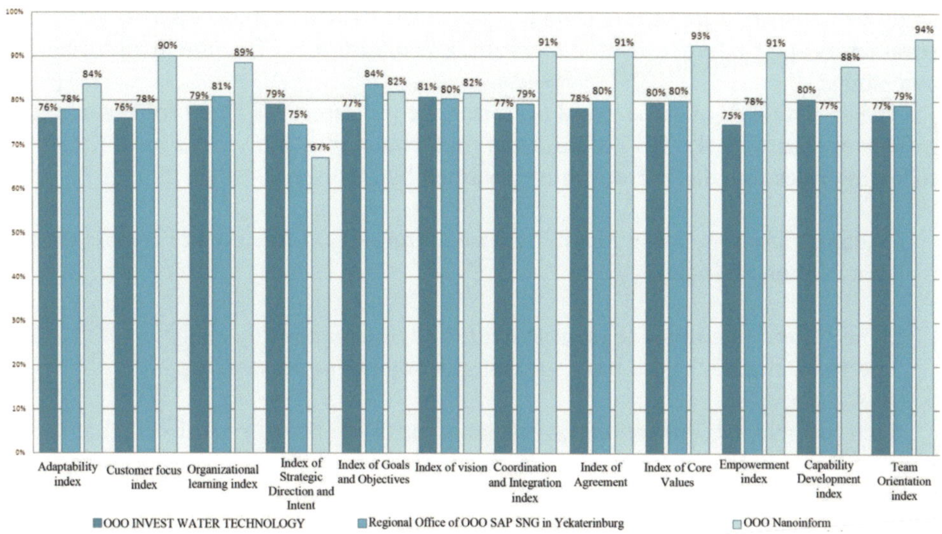

Figure 8.
Comparison of results for the Denison model indices (across the selected organizations).

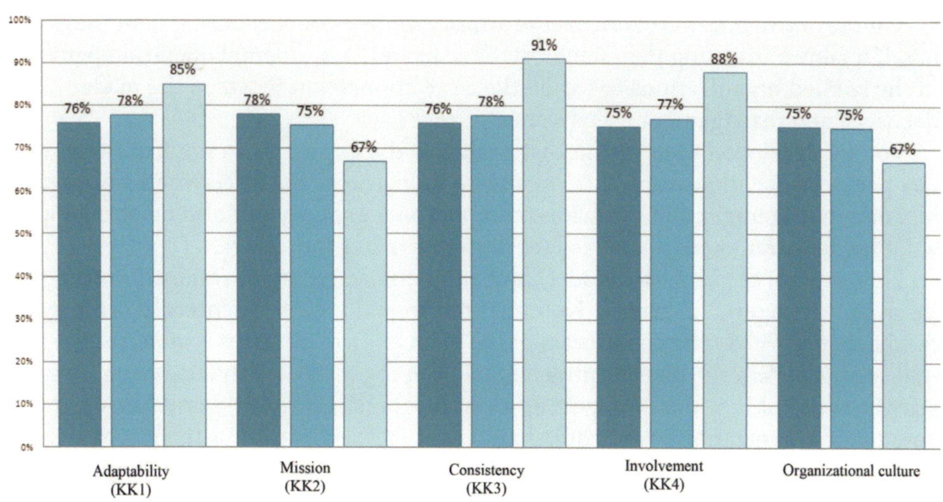

Figure 9.
Comparison of results for the final Denison model components (across the selected organizations).

They have a clear goal from which specific strategic objectives are built, thus ensuring a high degree of their involvement in the business process. Each employee feels involved in making any important decisions in the course of business that pursue a single common goal: to conduct business as efficiently as possible and to achieve a high level of income for the company in general and for each employee personally. A high degree of adaptability, which also plays an important role in shaping the organizational culture, is manifested in relationships with customers and employees and in mission understanding. The management of the surveyed organizations that position themselves as innovative prepares the employees of all levels to make decisions, to predict risks, and to be responsible for the outcome to some extent.

Therefore, for OOO Invest Water Technology, the main strengths are its mission understanding and consistency. For the Regional Office of OOO SAP SNG in Yekaterinburg, it is the increased adaptability and consistency, and for OOO nanoinform it is the consistency and staff involvement in business processes. But in general, the study demonstrated that there is no big gap in the values of indices across all the parameters and that organizational culture is economically effective in all three companies.

The causal fields of factors and performance indicators obtained for the organizations operating in the field of information technologies allow us to formulate the following rules that differ from those obtained by Denison:

1. Adaptability together with an increase in innovation has a major impact on net income and sales.

2. The system of employee involvement that determines the level of performance discipline has a great impact on sales and services.

Such formalization makes it possible to identify latent indirect effects that the implicit factor has on the key performance indicators of an organization using a mathematical tool, i.e., the fuzzy binary correspondences. The study revealed the most significant indirect effects. They can be interpreted as follows: "Implicit factor (organizational culture) has a significant effect on degree of innovation and on the quality of goods and services (influence coefficient 0.9), which in turn affect the increase in net profit and sales volumes."

On the one hand, this confirms the already known conclusions, and on the other hand, it allows adjusting the mental, statistical, and instrumental operation patterns in the studied organizations based on the close connection between the made decisions and management effectiveness.

So, the developed mechanism that evaluates the impact of an implicit factor on key performance indicators of an organization increases the effectiveness of organization management and evaluates them not only expertly but also by obtaining valid model parameters for the values of performance indicators.

Now we can see how our model can help the selected organizations to make informed managerial decisions. Note that traditional balanced scorecard offers no mechanism to verify the correctness of the quantitative estimates established for the indicator values in the planes (projections). The organization's management only sets the threshold estimates in getting a certain threshold result. For example, the amount of organization's net profit should be increased by no less than 24%, which is obviously set by experts, i.e., by managers of various levels by analyzing performance monitoring data over some period of time. A fragment of a strategic map in projection "finance" is considered, and the model-derived and raw information is summarized in **Tables 21–23**. Note that for each organization in question, a

Indicator designation	Indicator value before taking measures aimed at improving the implicit factor, RUR	Indicator value after taking measures aimed at improving the indicators of the implicit factor under the proposed model, RUR	Change in the indicator value, %	Initial indicator value according to the organization's strategic goal set when building a strategic map
OP_1	33,200	51,300	55%	Not less than 24%
OP_2	102,700	102,700	0%	Over 13%
OP_3	13%	18%	5%	Growth 15%
Composite indicator	80,926	85,494	6%	

Table 21.
Fragment of a strategic map for OOO nanoinform.

Indicator designation	Indicator value before taking measures aimed at improving the implicit factor, RUR	Indicator value after taking measures aimed at improving the indicators of the implicit factor under the proposed model, RUR	Change in the indicator value, %	Initial indicator value according to the organization's strategic goal set when building a strategic map
OP_1	3,230,000	4,522,000	40%	Not less than 24%
OP_2	6,458,000	6,458,000	0%	Over 13%
OP_3	6%	15%	9%	Growth 15%
Composite indicator	5,061,973	5,490,329	8%	

Table 22.
Fragment of a strategic map for OOO Invest Water Technology.

Indicator designation	Indicator value before taking measures aimed at improving the implicit factor, RUR	Indicator value after taking measures aimed at improving the indicators of the implicit factor under the proposed model, RUR	Change in the indicator value, %	Initial indicator value according to the organization's strategic goal set when building a strategic map
OP_1	4,530,000	5,546,428	22%	Not less than 24%
OP_2	7,378,000	7,378,000	0%	Over 13%
OP_3	10%	16%	6%	Growth 15%
Composite indicator	5,898,078	6,292,497	7%	

Table 23.
Fragment of a strategic map for the Regional Office of OOO SAP SNG in Yekaterinburg.

composite indicator calculated using the Mamdani fuzzy inference algorithm was introduced in projection "finance" in the form of a defuzzified value to characterize the overall change in situation for this projection after taking measures to achieve the strategic goals.

The calculated data obtained from the model can be interpreted by correlating it with the goal thresholds set in the strategic map of OOO nanoinform:

1. In general, after taking measures to improve the implicit factor (organizational culture), the composite indicator in projection "finance" increased by 6% indicating a positive dynamic.

2. Net profit increased by 55%, which is significantly higher than the expertly defined 24%; i.e., the goal is in fact achieved.

3. The maximum expenditure limit for the measures taken to improve the implicit factor will in this case be no more than RUR 18,100.

4. To improve other indicators in projection "finance," a number of other management activities are necessary to increase the value of selected indicators in the strategic map to an acceptable level. For example, profitability was increased by 5%, which fails to meet the 15% target level.

The calculated data obtained from the model can be interpreted by correlating it with the goal thresholds set in the strategic map of OOO Invest Water Technology:

1. In general, after taking measures to improve the implicit factor (organizational culture), the composite indicator in projection "finance" increased by 8% indicating a positive dynamic.

2. Net profit increased by 40%, which is significantly higher than the expertly defined 24%, i.e., the goal is in fact achieved.

3. The maximum expenditure limit for the measures taken to improve the implicit factor will in this case be no more than RUR 1,292,000.

4. To improve other indicators in projection "finance," a number of other management activities are necessary to increase the value of selected indicators in the strategic map to an acceptable level. For example, profitability was increased by 9%, which fails to meet the 15% target level.

The calculated data obtained from the model can be interpreted by correlating it with the goal thresholds set in the strategic map of the Regional Office of OOO SAP SNG in Yekaterinburg:

1. In general, after taking the measures to improve the implicit factor (organizational culture), the composite indicator in projection "finance" increased by 7% indicating a positive dynamic.

2. Net profit grew by 22%, which is close to the expertly defined 24%; however, the taken measures appeared to be insufficient, and other factors need to be identified in the organization's performance that would help achieve the set strategic goal. At the same time, this means that the organization in question has a fairly high level of organizational culture and should not spend money on boosting it at this stage.

3. The maximum expenditure limit for the measures taken to improve the implicit factor will in this case be no more than RUR 1,016,428.

4. To improve other indicators in projection "finance," a number of other management activities are necessary to increase the value of selected indicators in the strategic map to an acceptable level. For example, profitability was increased by 6%, which fails to meet the 15% target level.

The analysis of the obtained results and their interpretation according to the principles outlined above constitutes the final stage of the organization management process at which real actions are performed and plans are adjusted. At this stage, monitoring of the developed indicator panel plays a key role as they warn the management of potential problems and provide it with additional details and recommendations that facilitate the making of fast and adequate managerial decisions.

The developed management and managerial decision-making mechanism proved that the change (improvement or degradation) in the implicit factor (organizational culture) that was initially overlooked by all managers as a factor conductive to improving key performance indicators does in fact influence the organizational performance. At the same time, the proposed model offers a means to compare the level of expenditure on implicit factor transformation with the expected improvements in specific performance indicators and, therefore, in performance in general.

6. Conclusion

The methodology for making integrated evaluation of the impact exerted by the organizational culture based on fuzzy set descriptions has demonstrated:

1. The adequacy of our assumption that when presented as a linguistic variable, it requires that the temporal nature of its components be taken into account in that they can change over time (new ones added, old ones modified by assigning them new meaning).

2. The connection between the evolution of opinions in the organizational culture research with its implicit nature expressed in the direction in which its various components have been identified with varying degree of relevance depending on the period in the development of economic thought.

3. The multilevel and temporal nature of the effect that the organizational culture has on the organization's performance.

An instrumental analysis based on implementation of this methodology in the form of a web application allowed us to conduct multiple experiments and provided the statistical confirmation for our conclusions.

Under the proposed methodology for evaluating the organizational culture framed on the basis of the one proposed by Daniel Denison to evaluate the corporate culture of an organization, the main differences of the author's model are demonstrated and include the fuzzy inference implemented as a way to quantitatively compare and classify organizations by their level of organizational culture, which made this research less exclusive and more accessible not only for the large but also for the medium and especially small businesses.

In addition, the methodology described above implements:

1. A nonconventional interpretation of results in the study of organizational culture that identifies the levels of organizational culture as linguistic variable terms and addresses its causal character under the balanced scorecard methodology accepted in management theory and practice.

2. Not only the *qualitative* impact evaluation normally offered by various authors and scientific schools emphasizes the unconditioned direct relationship between the level of organizational culture development and the organization's

performance but also a *quantitative* one that answers the question of how much would a change in a given direction of organizational culture (under Denison's model) affect the key performance indicators of an economic entity, by the indirect effect evaluating technology that uses fuzzy binary correspondences and a fuzzy logical inference (according to Goguen).

3. Impact evaluation makes it more understandable for application in the organization management practice. In addition, this approach can strengthen the proposed model by distinguishing several levels of impact: a direct effect of various human resource management practices such as evaluation, career, training, and the resulting phenomena (involvement, loyalty, job satisfaction, turnover, etc.) on employment behavior, the effect on labor productivity of individual employees, and, finally, the effect on the overall performance of an organization and, hence, on its bottom-line performance.

Such approach may make the business interested in the organizational culture research as it can predict economic performance directly. By economic performance here, we mean the financial component of any business improvement, as well as a number of nonfinancial indicators improved incidentally. For science, this model is of interest primarily due to the fact that instead of organizational culture, it can work with any other implicit factor of an organization's performance that may not even be currently recognized by the theory and practice of an organization's business processes management.

Author details

Nazarov Dmitry Mikhailovich
Business Informatics Department, Ural State University of Economics, Yekaterinburg, Russia

*Address all correspondence to: slup2005@mail.ru

References

[1] Chatman JA, Barsade SG. Personality, organizational culture, and cooperation: Evidence from a business simulation. Administrative Science Quarterly. 1995;**40**:423-443

[2] Collins JC, Porras JI. Built to Last. New York: Harper Collins; 1994

[3] Dawid AP. The Decision-Theoretic Approach to Causal Inference. In: Berzuini et al., editor. Chapter 4. Oxford: Hart Publishing; 2012, pp. 25-42

[4] Deal T, Kennedy A. Corporate Cultures. Reading, MA: Addison-Wesleys; 1982

[5] Babbie ER. The Practice of Social Research. 8th ed. Balmont, CA: Wadsworth Publishing; 1998

[6] Harris PR. Management in Transition. San Francisco: Jossey-Bass; 1985

[7] Holmes S, Marsden S. An exploration of the espoused organizational cultures of public accounting firms. Accounting Horizons. 1996;**10**(3):26-53

[8] Introna L. Towards a Theory of Management Information. Electronic source. 1993. Available from: http://aisel.aisnet.org/cgi/viewcontent.cgi?article=1133&context=pacis1993

[9] Jaskyte K. Transformational leadership, organizational culture, and innovativeness in nonprofit organizations. Nonprofit Management & Leadership. 2004;**15**(2)

[10] Hair JF. A Primer on Partial Least Squares Structural Equation Modeling (PLS-SEM) [Text]. London, Copyright, 2014. 56 p

[11] Kotter JP, Heskett JL. Corporate Culture and Performance. New York: The Free Press; 1992

[12] Mamdani EH. Application of fuzzy logic to approximate reasoning using linguistic synthesis. IEEE Transactions on Computers. 1977;**26**(12):1182-1191

[13] Pearl J. Causality: Models, Reasoning and Inference. 2nd ed. Cambridge: Cambridge University Press; 2009

[14] Roy B. Multicriteria Methodology for Decision Aiding. Dordrecht: Kluwer Academic Publishers; 1996

[15] Rubin D. Bayesian Inference for Causal Effects: The Role of Randomization. The Annals of Statistics. 1978;**6**:34-58

[16] Wright S. Correlation and Causation. Journal of Agricultural Research. 1921;**20**:557

[17] Zadeh LA. Fuzzy sets. Information and Control. 1965;**8**:338-353

[18] Organizational culture diagnostics. Electronic source. Available from: http://www.eg-online.ru/article/78569/

[19] Ivanova TB, Zhuravleva EA. In: Ivanova TB, Zhuravleva EA, editors. Organizational Culture and Performance: Monograph. M.: RUDN; 2011. 152 pages.: illustrated

[20] Kaplan R, Norton D. The Balanced Scorecard: Translating Strategy Into Action. M.: Olimp-Biznes; 2003

[21] Kaplan RS, Norton DP. The Balanced Scorecard: Translating Strategy Into Action. 2nd ed. M.: ZAO Olimp-Biznes; 2005. Revised and extended. Translated from English

[22] Maister D. Practice What You Preach: What Managers Must Do To Create A High-Achievement Culture. M.: Alpina Biznes Buks; 2005. 246 pages. Translated from English

[23] Nazarov DM. Synergistic nature of the organizational work culture management as a business process of an industrial enterprise. In: Nazarov DM, Davydkin EV, editors. Ekonomicheskiy analiz: teoriya i praktika (Economic Analysis: Theory and Practice). 2012. No. 37. pp. 15-20

[24] Nasledov A. IBM SPSS Statistics 20 and AMOS: Professional Statistical Data Analysis. St. Petersburg: Piter; 2013. 416 pages

[25] Pervakova EE. On the Mechanisms of the Organizational Culture Influence on Performance and Productivity. Available from: http://www.civisbook.ru/files/File/Pervakova.pdf

[26] Cooke RA, Rousseau DM. Behavioral norms and expectations: A quantitative approach to the evaluation of organizational culture. Group & Organization Studies. 1988;**13**(3): 245-273

[27] Philosophical Encyclopedic Dictionary. 2nd ed. M.: Soviet Encyclopedia; 1989. p. 840

[28] Asher H. Causal Modeling. Sage; 1983

Section 8

Field-Programmable Gate Array for Fuzzy Controllers

Chapter 8

Functional Safety of FPGA Fuzzy Logic Controller

Mohammed Bsiss and Amami Benaissa

Abstract

In this paper we describe a methodology to implement a fuzzy logic controller in FPGA. The implementation of fuzzy logic controller (FLC) in FPGA requires a qualitative and a quantitative analysis to define the system safety integrity level (SIL). This level can be defined by the quantification of the probability of failure on demand (PFDavg). We propose to analyze the implementation advance safety architecture of fuzzy logic controllers with 1-out-of-2 controllers (1oo2) in FPGA using the reliability block diagram (RBD) and the Markov model. We demonstrate how from hardware characteristics parameters, such as rate of dangerous detected failure and undetected failure, the diagnostic coverage, proof test interval and other parameters to evaluate the PFDavg.

Keywords: fuzzy logic controller, safety integrity level (SIL), mean time to failure (MTTF), safe failure fraction (SFF), reliability block diagram (RBD), Markov model, average probability of dangerous failure on demand (PFDavg), field programmable gate array (FPGA), IEC standard 61508

1. Introduction

A synthesize fuzzy logic controller in field programmable gate array FPGA means that the VHDL code writing for the systems will be translated into gate, multiplexer, registers RAM, etc. Very low-level FPGA faults to high-level system hazards and common cause faults can put the FPGA-based systems in a dangerous state [1].

However, safety-related issues for FPGA-based systems remain to be not only verified but also following a safe methodology to design, implementation and evaluation such systems.

According to [2] the FPGA chip is classified as a type B with very complex structure. The first step was to perform failure modes, effects, and diagnostic analysis (FMEDA) for the safety related FPGA-based fuzzy logic controller.

FMEDA is a systematic process used in the development stage of an integrated circuit to ensure that it meets the pre-determined safety requirements. In the FMEDA, each component implemented in our FPGA is analyzed for possible failures and the consequences of these failures on the system.

The design, implementation and evaluation of a fuzzy logic controller in the field programmable gate array require a qualitative and quantitative analysis according to IEC 61508. Due to their usage in critical applications, the FLC have a very stringent average probability of failure on demand (PFDavg) requirement.

This requirement is usually determined by industry standards, such as the safety integrity level (SIL) rankings. The SIL is defined as a relative level of risk reduction provides by a safety function for safety function our FPGA-based FLC.

The safety function performed by the FLC maintains a safe state of the system relative to specific hazardous failures.

The four levels used in IEC 61508 are defined in **Table 1** [5] for various fractions of failures leading to a safe state as follows:

Safety integrity level	Probably of failure on demand
SIL4	10^{-4} to 10^{-5}
SIL3	10^{-3} to 10^{-4}
SIL2	10^{-2} to 10^{-3}
SIL1	10^{-1} to 10^{-2}

Table 1.
Definition of SILs for low demand mode from IEC 61508-1.

2. Definition and assumptions

2.1 Definition

Presented below is a glossary of terminology on topics related to functional safety used in this paper.

Diagnostic coverage represents the probability of discovering a failure. Diagnostic coverage of the test according to the safety standard Norm IEC 61508 is defined as the ratio of the rate of detected dangerous failures (by a diagnostic test) on the total failure of detected and undetected dangerous failure.

Safe failure fraction is used for calculating safety integrity levels (SIL).

Mean time to failure is the average time to the first failure.

Mean time between failure (MTTR) is time between two failures.

Probability of failure on demand (PFD) is a probability on the time interval that the system could not perform the function of safety for which it was at the time or the application of this function is made.

The safe failure fraction is defined by the ratio of average failures of safe λ_S plus dangerous detected failures λ_{DD} and safe plus dangerous detected and undetected λ_{DU} failures. The calculation is based on the architecture of FLC and on a functional analysis by carrying out a Failure Modes Effects and Diagnostic Analysis (FMEDA).

Safety integrity level (SIL) – Given a SIL to a system is a decision to be taken in consequence of process hazard and risk analysis. SIL defines the probability of dangerous failure that a system can be authorized. There are four possibility levels (SIL1, SIL2, SIL3 and SIL4) defined by safety norm IEC 61508 [2].

Component type A. All failure modes are known and can be detected. The value of the security factor S for components of type A on a worst-case is defined as S = 10% [3].

Component type B. All failure modes are not completely known. The value of the security factor S for components of type B on a worst-case is defined as S = 50%.

Proof test T-proof is periodic tests offline directed to detect failures in a system so that the system can be repaired to return in a state equivalent to its initial state.

Diagnostic tests are online test to detect hazardous failure. The diagnostic tests have an in fluent at the level of component (internal) but not at the level of the function of the security. The watchdog Test, Walking Bit Test and Ram Test are some example of diagnostic test.

Common mode failure refers to the simultaneous failure that can appear in two or more channels in a system multiple channels. The introduction of common-mode failures is generally represented by a factor of β. The 61,508 standard distinguishes two types of factor for non-detected dangerous failures and detected dangerous failures. The values for the factors Beta and are generally between 0.5% and 5% [4].

1oo2 architecture (one-out-of-two) consists of two channels perform each security function. The security function is executed once a channel request. Only any dangerous failure will lead to the failure of the function of application of both channels to lead to the failure of the security function on demand.

Reliability block diagram is a safety analysis for SIL selection for estimating the performance of systems, other methods are fault tree [6] analysis and Markov diagrams [7].

2.2 Assumptions

The technique and results developed in this paper are based on the assumptions following:

- Component failure and repair rate is assumed a constant failure over the life of the system.
- The hardware failure rates used as inputs to the calculations for a single channel of the subsystem
- All channels in a voted group have the same failure rate and diagnostic coverage rate.
- The proof test interval is at least one order of magnitude greater than the diagnostic test interval
- The demand rate and expected interval between demands are not considered.
- For each component of the safety system, the PFDavg is calculated, for simplification only from the undetected dangerous failure rate, λ_{DU} given in **Table 3** and the proof test interval, Ti.

Other assumptions can be referred to the Annex B of IEC 61508-6.

3. Architecture of fuzzy logic controller

For a simple architecture 1 out of 1 (1oo1), the fuzzy logic controller (FLC) contains a fuzzification process to change a real scalar input value to fuzzy value, a fuzzy inference engine for rule based expert systems and defuzzification to change fuzzy value into real scalar output. **Figure 1** presents the basis block diagram of simple fuzzy logic controller.

The parameter characterizing the present FLC are summarized in **Table 2**.

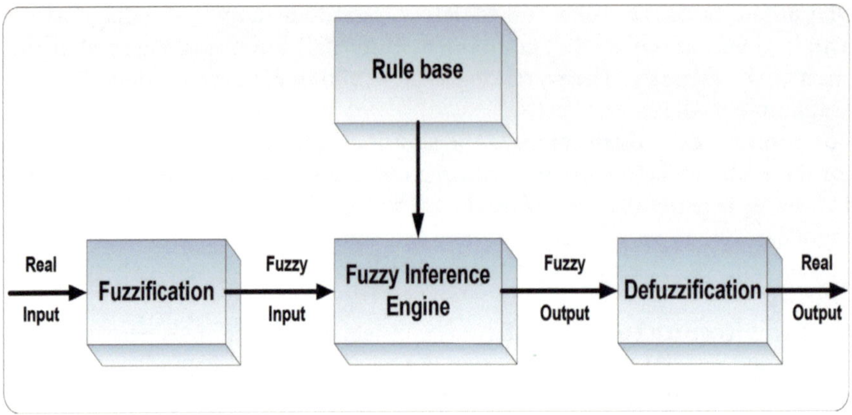

Figure 1.
Basis block diagram of simple fuzzy logic controller.

Fuzzy inference system	
Inputs	2
Outputs	1
Outputs resolution	12 bits
Antecedent MF's	7 trapezoidal
Antecedent MF resolution	14 bits
Consequent MF's	3 singleton
Antecedent MF resolution	12 bits
Aggregation method	Mandani Min-Max
Implication method	Product operator
Defuzzification	Weighted average

Table 2.
The parameter characterizing FLC.

The FLC has two inputs, one with four linguistic terms and the other with three and an output with three linguistic terms. This makes a total of 4 × 3 × 3 different rules that may be sued to describe the strategy of total control (**Figure 2**).

The FPGA-based fuzzy logic controller consists of two fuzzy logic controller (FLC) with the fuzzification process; rule evaluation process and defuzzification process in a redundant architecture (**Figure 3**).

In this kind of redundancy, the failure of one channel does not prevent the execution of the safety function. This architecture will be in dangerous state when both FLC have dangerous failures. The main advantage of this architecture is his low probability of failure on demand. Each FLC has diagnostic tests and the results of both FLC are controlled by the comparison module (**Figure 3**).

The safety function performed by the FLC maintains a safe state of the system relative to specific hazardous failures. The safety function is therefore the power loss for the analog outputs (de-energize-to-trip) of the system in case of dangerous failures by on-line diagnostics tests. These failures can be interconnect faults, stuck-at-fault, transition faults, the clock phase shift or a deviation of the value obtained respectively from the both controller.

Figure 3 shows a basic model for a fuzzy logic controller with redundancy architecture 1oo2 designed in FPGA.

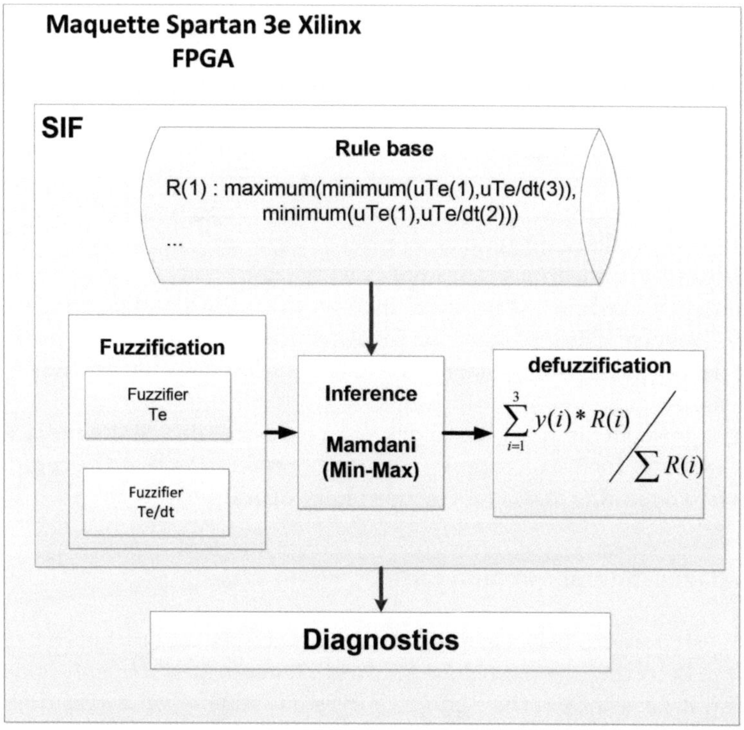

Figure 2.
Design of the present implemented FLC on FPGA.

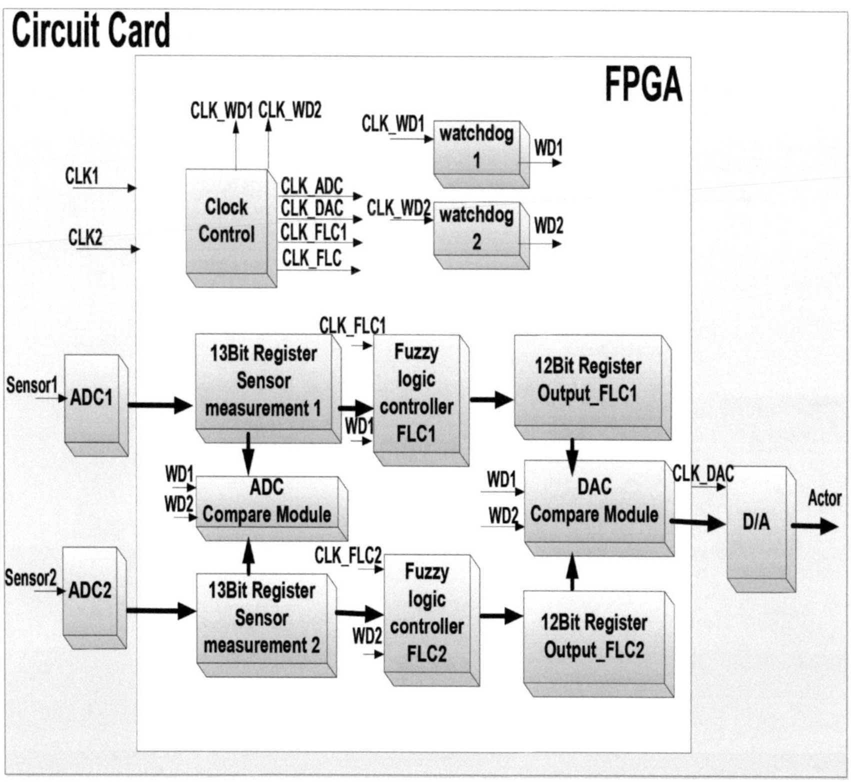

Figure 3.
Block diagram of the fuzzy logic controller with 1oo2 structure.

4. RBD and Markov model for safety integrity verification

4.1 Reliability block diagram

The reliability block diagram is a graphical representation of the system. Each component is represented by a function block in accordance with their logical relation of reliability (**Figure 4**). A series connection represent logic "and" of component and parallel connections represents logic "or", even as combination of series and parallel connections represents voting logic.

If a component fails in a series combination, the corresponding connection will be cut off. Conversely, in a parallel combination, the operation of a single instance is sufficient for the passage of the signal. System shutdown is only possible if all parallel instances fail.

Figure 4 presents the reliability block diagram associated to the fuzzy logic controller with the 1oo2 structure. We take in consideration that the components have only two operating states (correct or faulty operation).

4.2 Auto diagnostic and common cause

The first step was to perform a failure modes, effects, and diagnostic analysis (FMEDA) to detecting the hazardous hardware failures of systems. A failure is called safe if it does not put the FLC in a dangerous state when a hazardous fault occurs. A dangerous failure puts the logic controller in a potentially dangerous state and makes the system inoperative.

They are failure rates partitioned into four categories:

- Safe failure rate λ_s - do not have the potential to put the system in an hazardous state and is equal to the sum of safe detected failure rates λ_{SD} and safe undetected failure rate λ_{SU}

- Dangerous failure rate λ_D - have the potential to put the system in an hazardous state and is equal to the sum of dangerous detected failure rates λ_{DD} and dangerous undetected failure rate λ_{DU}

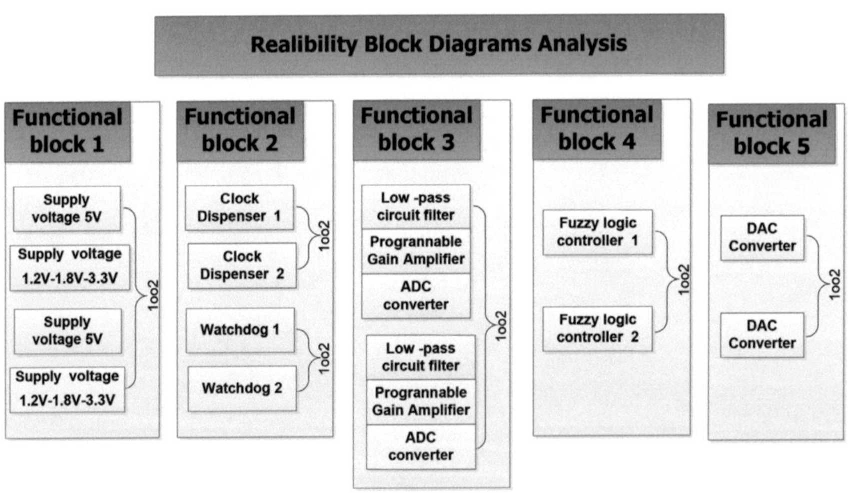

Figure 4.
Reliability block diagrams analysis.

- Dangerous detected failure λ_{DD} - is detected by the on-line diagnostics tests and the system will be placed into safe state.

- Dangerous undetected failure λ_{DU} - is undetected by on-line diagnostics tests and the system will not be placed into safe state.

By redundancy systems the combination of on-line diagnostic and common-cause was included. Since the failure is partitioned into eight categories [7].

- Safe, detected normal λ_{SDN}
- Safe, detected, common-cause λ_{SDC}
- Safe, undetected normal λ_{SUN}
- Safe, undetected common cause λ_{SUC}
- Dangerous, detected normal λ_{DDN}
- Dangerous, detected, common-cause λ_{DDC}
- Dangerous, undetected normal λ_{DUN}
- Dangerous, undetected common cause λ_{DUC}

The possible failures of the fuzzy inference engine implemented in FPGA and their classification are presented in **Table 3**.

Type of failure	Potential causes	Diagnostic test	Classification of failure
Hazardous hardware failure in module fuzzification	Stuck-at Low or Stuck-at High anomaly at the internal FPGA component	Periodic comparison of the result of the redundancy controllers.	Dangerous detected Failure λ_{DD}
Hazardous hardware failure in module inference rule			
Hazardous hardware failure in module defuzzification			
Failure of an internal element that does not intervene in the logic implemented in FPGA	Stuck-at Low or Stuck-at High anomaly at the internal FPGA component	No diagnostic	Since it does not affect the security function of the FLC then it is an undetected safe failure λ_{SU}
Flash memory failure where logic (VHDL code) is stored.	Hardware fault, electrostatic disturbance, magnetic waves, high voltage frequencies, etc.	Examining of cyclic redundancy value	A failure in the flash memory during FLC operation can be detected only after the mission time delay Ti. It can therefore be classified as a detected safe failure λ_{SD}
The drift of the clock		Examining via eatchdog circuit	Dangerous detected Failure λ_{DD}

Table 3.
Failure mode distribution for functional block 3 (FLC).

4.3 Quantitative analysis using RBD

The structure of reliability block diagram (RBD) defines the logical interactions of failures within a fuzzy logic controller implemented in FPGA. Each component of the fuzzy logic controller is a functional block connected by a series for output module DAC and parallel structure for measurement units. **Figure 4** presents the reliability block diagram associated to each component. The unreliability data for each subsystem components is given in **Table 4**. The probability PFDavg is calculated by summing the probability of failure of all the functional blocks of a FLC. The quantification of average frequency of dangerous failure of our safety function is giving by Eq. [8]:

$$PFDavg = 2\left((1-\beta_D)\lambda^{DD} + (1-\beta)\lambda^{DU}\right)^2 t_{CE} t_{GE} + \beta_D \lambda^{DD} MTTR + \beta \lambda^{DU}\left(\frac{T_i}{2} + MTTR\right) \tag{1}$$

The time of unavailability of a channel t_{CE} due to a detected dangerous failure is given by the following formula [8]:

$$t_{CE} = \frac{\lambda^{DU}}{\lambda^{D}}\left(\frac{T_i}{2} + MTTR\right) + \frac{\lambda^{DD}}{\lambda^{D}} MTTR \tag{2}$$

The time of unavailability of the other channel t_{GE} is also added because of detected dangerous failure which is represented by the following formula [8]:

$$t_{GE} = \frac{\lambda^{DU}}{\lambda^{D}}\left(\frac{T_i}{3} + MTTR\right) + \frac{\lambda^{DD}}{\lambda^{D}} MTTR \tag{3}$$

This result gives a PFDavg of 2.7426E−03, which corresponds to a safety integrity level of SIL2.

The subsystem PFDavg contribution for the supply voltage is 2.1920E−03, for the fuzzy controller implemented in FPGA is 7.3616E−06. That means that the on-line diagnostics tests implemented for FLC systems in FPGA is with high performance and efficiency (**Figure 5**).

4.4 Quantitative analysis using Markov model

Markov modeling brings a good reliability and safety techniques for qualitative and quantitative analysis that uses state diagrams. This method take account for a realistic repair time, probability of correct repair, proof test effectiveness, and automatic diagnostic testing. The Markov system model for redundancy structure

Component	HFT	PFDavg	% of total PFDavg	*SFF*
Supply power	1	2.1920E−03	82.93%	90.000%
Clock dispenser	1	1.8785E−04	6.84%	92.059%
ADC converter	1	2.1818E−04	7.95%	90.235%
Fuzzy controller	1	7.3616E−06	0.68%	99.500%
DAC controller	1	5.4770E−05	1.99%	95.000%
Total		**2.7426E−03**	**100%**	**92.57%**

Table 4.
Failure mode distribution and SIL performance analysis for FLC system.

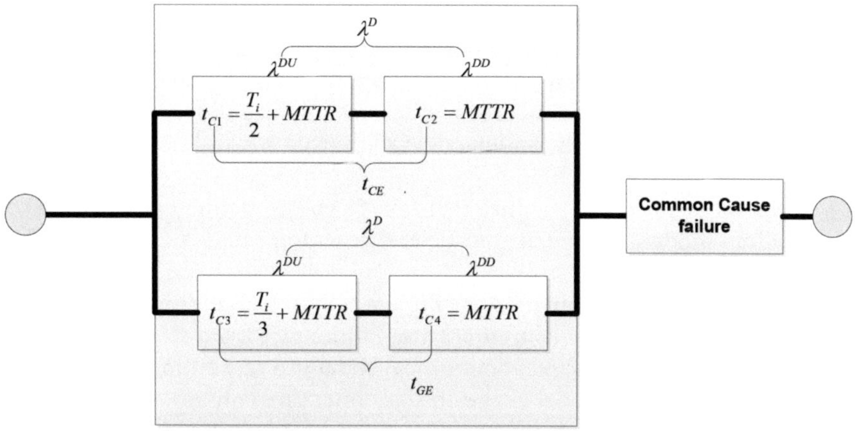

Figure 5.
Schematic design of the reliability principle (1oo2).

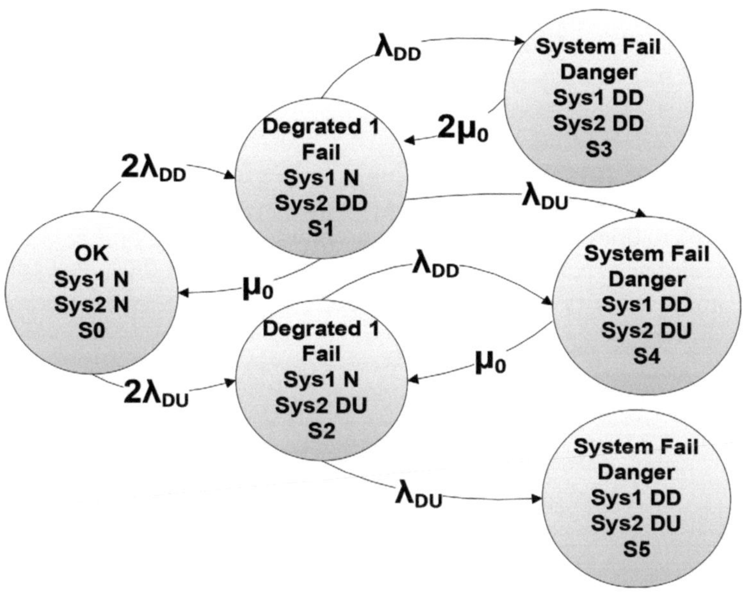

Figure 6.
Markov model of the 1oo2 architecture diagnostic (no common cause).

1oo2 with only on-line diagnostic is presented in **Figure 6**. This Markov model of the 1oo2 architecture contains 6 states [7]:

- The first state (S0): specifies the normal state where the booth controller properly works.
- The second state (S1): specifies the state where one controller of the system has a dangerous detected failure by diagnostic with transition probability of $2\lambda_{DD}$. The system can be repaired according to the transition rates μ_0.
- The third state (S2): specifies the state where one controller of the system has a dangerous undetected failure with transition probability of $2\lambda_{DU}$ and the second work properly.

- States (S3), (S4) and (S5): specify a system fail state, where the booth controllers have a dangerous detected failure by on-line diagnostics tests (S3), or one controller has a dangerous detected failure also by on-line diagnostics tests and the other has a dangerous undetected failure (S4), or the booth channels have a dangerous undetected failure (S5) by on-line diagnostics tests.

A Markov model of 1oo2 structure that take in consideration combination of different failure modes, on-line diagnostic and common cause is draw in **Figure 7** with six states [7].

It has the same state combinations as **Figure 6** with two additional failure lines. There is a dangerous detected common-cause failure rate from state (S0) to state (S3) and a dangerous undetected common-cause failure rate from state S0 directly to state (S5). The Markov model of the 1oo2 architecture contains 6 states, in that case the transition matrix P with dimension (6 × 6) is given by [7].

$$p = \begin{bmatrix} 1-(\lambda_{DC}+2\lambda_{DN}) & 2\lambda_{DDN} & 2\lambda_{DUN} & 2\lambda_{DDC} & 0 & \lambda_{DUC} \\ \mu_0 & 1-(\lambda_D+\mu_0) & 0 & \lambda_{DD} & \lambda_{DU} & 0 \\ 0 & 0 & 1-\lambda_D & 0 & \lambda_{DD} & \lambda_{DU} \\ 0 & 2\mu_0 & 0 & 1-2\mu_0 & 0 & 0 \\ 0 & 0 & \mu_0 & 0 & 1-\mu_0 & 0 \\ 0 & 0 & 0 & 0 & 0 & 1 \end{bmatrix} \tag{4}$$

The transition matrix P is a matrix showing the probabilities‘ distribution of different states in one time interval. This matrix can be multiplied by itself to get transition probabilities for different time intervals.

The FLC system is starting always by one particular state (S0), so it contains a single one and a quantity of zeros. The starting probability S would be:

$$S^0 = [1 \quad 0 \quad 0 \quad 0 \quad 0 \quad 0] \tag{5}$$

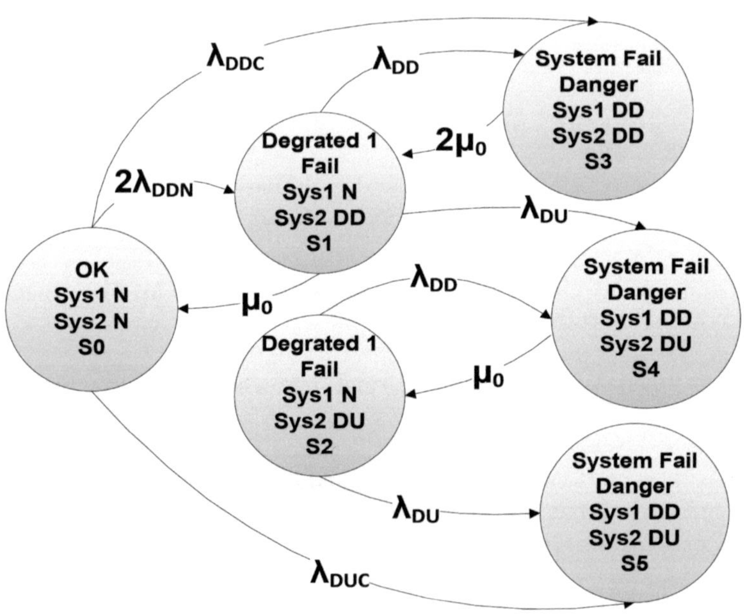

Figure 7.
Markov model of the 1oo2 architecture—diagnostic and common cause.

It means that the probability to be in normal state at initial time is 100 and 0% for the other states.

After 1 year, the system average frequency of dangerous failure of the safety function is the sum of the of all functional components probabilities of the 1oo2 FLC systems:

$$PFD_{avg} = \sum PFD_{avg\_Subsystem} \quad (6)$$

The FLC with 1oo2 structure is always starts in state zero. After n hours, the calculation process of the distribution probabilities S^n is:

$$S^1 = S^0 \times P$$
$$S^2 = S^1 \times P$$

This process can be continued as necessary.

$$S^3 = S^2 \times P$$
$$S^4 = S^3 \times P$$
$$\dots$$
$$S^n = S^{n-1} \times P$$

The S^n matrix for any particular time interval is obtained by multiplying S^{n-1} times P. This process can be continued as necessary, and the probability distribution increases progressively each time, then that remains unchanged as time progresses. If $S^{n+1} = S^n$ a limiting state probability is reached. This matrix is labeled P^L.

$$S^L = S^n \times P = S^{n-1} \times P$$

The FLC with 1oo2 structure has a safe failure rate of 6.6302E−07 failures per hour and a dangerous failure rate of 1.9118E−07 failures per hour. On-line diagnostic detect 95% of dangerous failure and 92% of safe failure. When failures are detected, the average system repair time is 24 hours.

The beta factor β is estimated to be 2%. The failure rates are divided by diagnostic capability. The following failure rates result:

$$\lambda^{SD} = \lambda^S \times 0.92$$
$$\lambda^{SU} = \lambda^S \times (1 - 0.92)$$
$$\lambda^{DD} = \lambda^D \times 0.95$$
$$\lambda^{DU} = \lambda^D \times (1 - 0.95)$$

These failure rates are multiplied by beta factor using following equations:

$$\lambda^{SDN} = (1 - \beta) \times \lambda^{SD}$$
$$\lambda^{SDC} = \beta \times \lambda^{SD}$$
$$\lambda^{SUN} = (1 - \beta) \times \lambda^{SU}$$
$$\lambda^{SUC} = \beta \times \lambda^{SU}$$
$$\lambda^{DDN} = (1 - \beta) \times \lambda^{DD}$$
$$\lambda^{DDC} = \beta \times \lambda^{DD}$$
$$\lambda^{DUN} = (1 - \beta) \times \lambda^{DU}$$
$$\lambda^{DUC} = \beta \times \lambda^{DU}$$

Where the failure rates and repair rates are substituted into the transition matrix P, the following solving for limiting state probabilities, the results are:

$$S_0^L = 0.9583$$
$$S_1^L = 0.0095$$
$$S_2^L = 0.0095$$
$$S_3^L = 0.0093$$
$$S_4^L = 0.0097$$
$$S_5^L = 0.0038$$

Since the system is down (failed) in state 5, the predicted average steady-state downtime is 0.0038. The control system is successful in state S0, S1 and S2; therefore, we add the limiting state probability of the success states equal to 0.9773%.

5. Conclusion

Markov analysis is used to analyze different states that take the system during its life cycle. Markov analysis provides information on the probability of FLC.

This application contains several important assumptions. First, notice that in Markov models M-out-of-N the probabilities in each row sum to one. Second, the probabilities in Markov models will not change over time. Third, the states are independent over time. In a Markov process after a number of periods (500 hours) have passed, the probability will approach steady state. For our example, the steady-state probabilities are:

- 13.5E−3 per hour = probability of the FLC to be in a dangerous undetected failure.

- 1.9E−2 per hour = probability of FLC degraded system fail.

However, the reliability block diagram analysis is based on the IEC 61508 international standard in the calculation of PFDavg. This standard considers all the parameters defined previously and there is a difference between both type components A and B. The type of components allows identifying the safety factor which contributes directly in the calculation of the PFDavg. Despite this difference between both standards, both analysis methods give the same results.

The FLC with redundancy structure 1oo2 has a redundant architecture with two controllers adopted by the FLC and the watchdog. This architecture has a majority voting arrangement for the output signals. If only one FLC gives a result which disagrees with the other FLCs, the output state does not change.

The probability of FLC with 1oo2 architecture to be in a dangerous undetected failure is 2.7426E−03 per hour, which relocates the system safety integrity level to SIL2.

List of abbreviations

ADC	analog digital converter
FMEDA	failure modes, effects, and diagnostic analysis
RBD	reliability block diagram
IEC	International Electrotechnical Commission

DAC	digital analog converter
DC	diagnostic coverage
E/E/PE	system electric, electronic, electronic programmable
FPGA	field programmable gate array
ISO	International Organization for Standardization
FLC	fuzzy logic controller
MTBF	mean time between failures
MTTF	mean time to failure
MTTR	mean time to repair
MooN	a system of N redundant channels has a M-out-of-N voting
PFD	probability of failure on demand
PFDavg	average probability of failure on demand
PFH	probability of a dangerous failure per hour
SFF	safe failure fraction
SIF	safety instrumented function
SIL	safety integrity level
SIS	safety instrumented system
VHDL	very high speed integrated circuit hardware description language

Author details

Mohammed Bsiss* and Amami Benaissa
Department of Computer Science, Systems and Telecommunications (LIST), Faculty of Science and Technology, Tangier, Morocco

*Address all correspondence to: fstbsisss@gmail.com

References

[1] IEC. 61508-6: Functional safety of electrical/electronic/programmable electronic safety-related systems. e2.0d; 2010

[2] IEC. 61508-2:2010: Functional safety of electrical/electronic/programmable electronic safety-related systems (E/E/PE, or E/E/PES). e2.0d, pp. 27, Table 3

[3] IEC. 61508-2:2010: Functional safety of electrical/electronic/programmable electronic safety-related systems (E/E/PE, or E/E/PES). e2.0d, pp. 77

[4] IEC. 61508-6:2010: Functional safety of electrical/electronic/programmable electronic safety-related systems (E/E/PE, or E/E/PES). e2.0d, pp. 92, Table D.4

[5] IEC. 61508-2:2010: Functional safety of electrical/electronic/programmable electronic safety-related systems (E/E/PE, or E/E/PES). e2.0d, pp. 34, Table 3

[6] ISA TR84.0.0.2. Safety Instrumented Functions (SIF), Safety Integrity Level (SIL), Evaluation Techniques. Part 2: Determining the SIL of SIF Via Simplified Equations. North Carolina; 1998

[7] Goble LWM. Control Systems Safety Evaluation and Reliability. 3rd ed. Research Triangle Park, NC: International Society of Automation; 2010

[8] IEC. 61508-6:2010: Functional safety of electrical/electronic/programmable electronic safety-related systems (E/E/PE, or E/E/PES). e2.0d, pp. 143-144